THE SCIENTIFIC ARTICLE IN THE AGE OF DIGITIZATION

Information Science and Knowledge Management

Volume 11

The titles published in this series are listed at the end of this volume.

THE SCIENTIFIC ARTICLE IN THE AGE OF DIGITIZATION

by

JOHN MACKENZIE OWEN
University of Amsterdam,
The Netherlands

Springer

A C.I.P. Catalogue record for this book is available from the Library of Congress.

ISBN-13 978-90-481-7346-4
ISBN-10 1-4020-5340-1 (e-book)
ISBN-13 978-1-4020-5340-5 (e-book)

Published by Springer,
P.O. Box 17, 3300 AA Dordrecht, The Netherlands.

www.springer.com

Printed on acid-free paper

Contents

List of Figures

List of Tables

Chapter 1

INTRODUCTION

1. The birth of the electronic journal

In the autumn of 1987 Michael Ehringhaus and Bird Stasz of Syracuse University launched *New horizons in adult education*, probably the very first refereed scientific journal[1] to be published in electronic form (fig. 1.1). The first issue was sent over the Adult Education Network (AEDNET). The journal still exists today.[2] In March 1991 Ted Jennings of the University at Albany (State University of New York) launched *EJournal*,[3] described as an 'electronic journal concerned with the implications of electronic networks and texts' (fig. 1.2), coining the now popular term *e-journal*. The *Online journal of current clinical trials,* published from September 1991, has been described as the first *peer reviewed* electronic journal in medicine.[4] [5]

How significant is the phenomenon of the scientific electronic journaltoday, more than a decade and a half after its introduction? Over this period information and communication technologies (ICT) have been an important factor in the development of scientific communication. Applications such as communication over digital networks (converging towards a single network – the Internet), the use of computerized systems for creating, storing and retrieving

[1] The scientific journal is also referred to as 'scholarly' or 'research' journal. We use the former term in this study.

[2] Although no longer distributed over AEDNET but through the World Wide Web. The first issue is now available at http://www.nova.edu/~aed/horizons/vol1n1.

[3] See http://www.ucalgary.ca/ejournal/archive/ej-1-1.txt for the inaugural edition.

[4] Henshaw 2001.

[5] An important source for the history of early e-journals is a survey by Hitchcock et al., (1996), who at the end of 1995 identified over 100 'online' journals in the domain of science, technology and medicine (STM), of which 35 were 'electronic only'. Another early overview is given by Roes (1994), who identified 39 refereed scientific electronic journals.

```
****************************************************************************
****************************************************************************
****************************                            ****************************
*********************                                          *********************
***************                                                      ***************
************                                                            ************
********                                                                    ********
******                                                                        ******
****                     NEW HORIZONS IN ADULT EDUCATION                        ****
***                                                                              ***
**                                                                                **
****************************************************************************
  Volume 1                        Number 1                        Fall 1987
****************************************************************************



EDITORS


Michael Ehringhaus.......Syracuse University

Bird Stasz...............Syracuse University



EDITORIAL BOARD


Michael Law..............University of British Columbia

Jane Hugo................Syracuse University

Tom Sudduth..............University of Wyoming

Hank Healy...............Cornell University

Judith Adrian............University of Wisconsin

Joyce Stalker Costin.....University of British Columbia

Priscilla Spencer........Columbia University


CONTENTS


Adult Education in Nicaragua: Adapting and Growing in a Changing Reality
.........................by Samuel Simpson............................
```

Figure 1.1. The first electronic journal: New horizons in adult education

scientific information, and the migration from print to digital formats are examples of developments that have impacted on the way scientists exchange information. The emergence of the electronic journal as a substitute for the printed scientific journal has been regarded by many as a significant innovation if not a 'revolution' in scientific communication. It is clear even to a superficial observer that digitization has changed the distribution mode of the scientific journal, but it is not at all clear what that means for the scientific

```
March 1991          _EJournal_  Volume 1 Issue 1          ISSN # 1054-1055

            An Electronic Journal concerned with the implications
                      of electronic networks and texts.

            University at Albany, State University of New York
                         ejournal@albnyvms.bitnet

                     There are 421 lines in this issue.

CONTENTS:

 Electronic Journals of Proposed Research              226 lines.
        by Robert K. Lindsay
           Mental Health Research Institute
           University of Michigan

DEPARTMENTS:
  Letters                                              11 lines.
  Reviews                                              10 lines.
  Supplements to previous texts                        12 lines.
  Pointers to texts appearing elsewhere                21 lines.
  Information about _EJournal_                         92 lines.

----------------------------------------------------------------------------
This electronic publication and its contents are (c) copyright 1991 by
_EJournal_.  Permission is hereby granted to give away the journal and its
contents, but no one may "own" it.  Any and all financial interest is herel
assigned to the acknowledged authors of individual texts.  This notificati‹
must accompany all distribution of _EJournal_.
----------------------------------------------------------------------------
```

Figure 1.2. The first issue of Ejournal

article itself. Although many claims have been made, e.g. about issues such as multimedia and interactivity, very little empirical research has been carried out in this area. This book is predicated on the idea that if the electronic scientific journal – as the outcome of a process of digitization – can make any claims to a revolutionary impact that has transformed scientific communication, this would have to be reflected in the genre of the scientific article itself. More specifically, we shall address the question how the ongoing process of digitization has impacted on the *substance* of formal scientific communication as reflected in the scientific article as a representation of research outcomes. This question is the main subject of this book.

Analyzing the impact of digitization on the substance of scientific communication is not merely an academic exercise. An understanding of the social system of scientific communication and its development over time is a prerequisite for designing and implementing policies and socio-technical changes that aim to improve the production and dissemination of scientific information. At the present moment, the predominant mechanism for achieving such improvements is the use of new information and communications technologies. But in order to implement new technologies effectively, we need not only to understand the communication system as such, but also how previous applications of technology have been taken up by the scientific community.

However, the dynamics of technological development are such that there is often little room for reflection. Technology is often seen as driving society almost of necessity towards innovation that is based more on the technically possible than on what has actually proved to be effective in social reality. As a result, information science is often preoccupied with predicting potential impacts of new technology and developing solutions that are believed to optimize its possibilities. Information science is therefore oriented more towards understanding the future than towards interpreting the past. In contrast, this study can be regarded as 'historical' in that it looks back at the way the larger societal process of digitization and the application of ICT have impacted on and resulted in the current state of scientific communication.

The justification for our approach here is to be found in the problematicrelationship between the *communication of research* and *digitization.* Since the Scientific Revolution it has been acknowledged that the difference between an individual's belief and scientific knowledge is that the latter is 'a public and shared commodity'.[6] For this to be the case – since not all researchers can be supposed to participate in the same research at the same time and place – a formal system of research communication is essential. Such a system has developed since the Philosophical Transactions and the Journal des Sçavans of 1665 into the high volume global system we know today.[7]

In our 'information society' the importance of ICT, including the ongoing digitization of information resources, is of course unquestionable. It is also clear that, after initial hesitation, the academic publishing world has adopted ICT on a wide scale - at least 17.000 scientific journals are now estimated to be available in digital form. However, it is still not at all clear how the digitization

[6] Shapin 1996, p. 106.

[7] Statistics are notoriously contentious in this area. But to give just one example: the number of biomedical journals increased from 22 in 1775 to 19.000 in 1973 Corning and Cummings 1976, quoted in Henderson 2002a. See also Henderson 2002b.

of the scientific journal has transformed the scientific article itself and/or to what extent the adoption of ICT in the practice of research is reflected in the way the research community reports on that research. More insight into these issues would contribute to policy making both in academia and in the publishing world. And, perhaps more importantly, it would satisfy our own curiosity to explain the tension between pretentious claims and expectations about the impact of digitization on scientific communication, and the impression that in fact there has been little fundamental change.

For the purpose of this study we define (formal) scientific communication as a process where certified scientific or scholarly findings and research outcomes are recorded on a durable communications medium with the *express purpose* of transferring them *over space and time* to other recipients. The primary *genre* for scientific communication is the *(research* or *scientific) article,* a form that has developed since the publication of the first scientific journals in the second half of the 17th century. The terminology used to describe the phenomena studied here is somewhat confused. We use the term 'scientific communication' only for the formal communication of research results. We do not discuss various informal modes of communication within the scientific community such as e-mail, pre-prints, 'grey' literature, etc., that are sometimes included in the concept of scientific communication. Another term used for this field is 'research communication'. Instead of 'scientific' the adjective 'scholarly' is also common. The words 'science' and 'scientist' are used here in a broad sense, encompassing (researchers in) all areas of academic enquiry: the humanities, the social sciences as well as what is often referred to as 'STM': science, technology and medicine or life sciences.

Outline of this chapter

In this introductory chapter we proceed as follows. First we describe common perceptions of the impact of digitization on scientific communication – as exemplified by the electronic journal – that are often informed by concepts such as 'revolution', 'transformation' and 'crisis'. We then discuss a number of current issues pertaining to more specifically to electronic journals. In subsequent sections we cover the information science context, theoretical considerations, methodological approach and level of analysis.

2. The electronic journal 'revolution'

It is often taken for granted that the information and communication technologies (ICT) that have come into use over the past decades have revolutionized scientific communication as they are said to have other areas of society. Already in 1970, Benedek wrote about the 'scientific-technical revolution' in the context of technical libraries, while a year later, Michel Menou gave a paper on 'Information revolution or revolution for information'.[8] The revolutionary connotation has never lost its attractiveness since. In 1991, Moulthrop referred to hypertext as a revolution. Harnad (1996) discusses peer review in the context of what he describes as 'the [Internet's] real revolutionary dimension: interactive publication in the form of open peer commentary on published and ongoing work'. Friend (1998) even suggests that 'the availability of print journals in an electronic format is only the beginning of a revolution in the communication of research'. Treloar (1999) writes that 'in the last decades of the twentieth century, technological developments have revolutionized our attitudes towards communication as well as our ability to communicate ideas and research results'. Hunter (2001) has written about content management as part of 'the electronic publishing revolution'. Eisend (2002) discusses the extent to which the Internet has 'revolutionized' academic research and publication. In 2000 the American Special Libraries Association held its annual conference on the theme 'Independence to interdependence: the next phase in the information revolution', a theme that was thought to 'recognize the dynamic and evolutionary nature of the phenomenon known as the 'information revolution' and look forward to the next stage in its development.' As a conference theme this was not new: the American Society for Information Science had already held its annual meeting in 1975 under the theme 'Information revolution'. Gorniak-Kocikowska (2001), in an article on 'Revolution and the library,' writes that 'in all likelihood, the computer revolution will have an even more profound impact on the library than did the printing press revolution'.[9] Even individual technical developments such as the 'semantic web' are presented as 'revolutions'.[10] Others, including Berners-Lee et al., (2001) and Sosteric et al., (2001) believe the revolution still has to come. Schaffner (1994) expressed the general feeling of the early phase in electronic publishing as follows:

> 'New technologies will soon bring fundamental changes to the process of scientific communication. ... It is clear to all involved – from the most astute students of electronic publishing to the most casual observers – that we are approaching a time when new information technologies will cause profound and elemental changes in scholarly communication. While these changes will eventually affect communication in all areas of scholarship, the sciences seem likely to be affected first.'

The idea that we are on the doorstep of a revolution in scientific communication has not subsided now that we are able to obtain a more balanced perspective on the impact of ICT. In 2003, in an overview of the history of the scientific article, Harmon and Gross describe the scientific article as:

[8] Menou 1971.
[9] For a comprehensive review of predictions and speculations regarding the future of academic libraries see Sapp and Gilmour 2002, 2003.
[10] Lu et al., 2002.

> 'in the midst of a radical transformation spurred by advances in computer technology ... the next century may well witness the extinction of the original scientific 'paper' appearing on paper. ... And the long-term effect of electronic preparation and publication of manuscripts may be as profound as when the scientific article evolved from scholarly letter writing and books in the seventeenth century'. [11]

The idea of an electronic revolution in scientific communication is related to ideas about the 'liberation from print' and the explosive effects of electronic text. Lanham argues that electronic text will 'disempower the force of linear print', Landow speaks of 'liberation' from the 'tyrannical ... voice of the novel', and Bolter argues that 'what is unnatural in print becomes natural in the electronic medium'.[12] However, it remains difficult to answer the question in what sense these assertions are true and how the current or future state of scientific communication really differs from before the advent of digitization.

The idea of a revolution in scientific communication also derives from a more widespread concept of a 'digital revolution' in modern society. Chien (1997) describes the convergence of computing, communications and content industries as 'one of the most dramatic changes in the ongoing information revolution'. Duff (1998) refers to the 'information revolution' in the context of Bell's economic theory of the information society, while Gunn (2000) of the Center for Applied Policy Research points to 'new digital societies coalescing around access to and utilization of personal communications, computational resources and electronic consumer goods, all of which are born out of the digital revolution'. Chodorow (1998) also refers to convergence – especially of media business and education – in the context of an 'electronic revolution'. In 2002 the Economist summarized it all for the average business manager in a book called 'E-trends: making sense of the electronic communications revolution'. Nwaobi (2001), another economist, refers to a 'transition ... described as digitalization (or knowledge revolution) driven by knowledge and by the technologies for processing and communicating it' that might even lead to 'a vision of a new society in which humans live in harmony with each other and with nature'. The utopian undertones of the revolutionary theme are often clear. Kling and Lamb (1996) have criticized such utopian visions for a lack of empirical analysis, adding that 'it is ironic that computing – often portrayed as an instrument of knowledge – is primarily the subject of discourse whose knowledge claims about its impact are most suspect'.

[11] Harmon and Gross 2003, session 5.

[12] Lanham 1994, p. 21, Landow 1992, p. 10-11 and Bolter 1991, p. 143, quoted in Duguid 1996 who provides further examples of accounts arguing for either replacement or liberation of the book by electronic media.

3. Electronic journals: the issues

The discourse about ICT and scientific communication, whether or not phrased in terms of a 'revolution', is often confined to the level of the communication system, involving publishers, libraries, and scientists in their roles of author and user. The major issues in the debate are *technical* (e.g. the functional solutions for scientific communication), *economical* (including the the future and economic viability of institutions in the information chain), *legal* (especially copyright) and *behavioral* (e.g. acceptance of digital formats by the scientific community). Much of the literature is missionary in nature, based on technological optimism and determinism, and on a sense of what will or at least *should* be the application of ICT and its impact on scientific communication.

But there is also a sense of urgency and crisis, reflected in concepts such as the 'serials crisis' and the 'publishing crisis'. Crowther (1999) describes this crisis as 'a crisis in communication, where library funding cannot keep pace with scientific output', caused by 'focusing on scholarly information as a commodity to be sold'. Abramson (2000) discusses the advantages and disadvantages of Web publishing as a response to the 'crisis in scientific communication'. This sense of crisis is often informed by a perceived lack of control over the process of scientific publishing, for which ICT might provide a solution.[13] But not everybody shares such optimism about the benefits of ICT: Solomon (2002) points to 'the [profound] implications of electronic distribution for ownership and access to the scholarly literature', adding that they are 'likely to exacerbate the already serious serial pricing crisis that is hindering the widespread access to scientific and scholarly information'.

In 1997 a Canadian Task Force on Academic Libraries and Scholarly Communication reported on such issues, arguing that 'failure to address the limitations of the current scientific communication system will have serious repercussions for Canada's knowledge enterprise'.[14] Halliday and Oppenheim (2001) assert that 'digital publishing by publishers has done nothing to relieve the problem [that] prices have consistently increased annually at a rate well above the general inflation rate for the last two decades'. The demise of the scientific monograph is also sometimes seen as an emerging problem.[15] In a Delphi survey carried out by Keller (2001) respondents were found to believe that 'throughout their 300-year history, journals have never been faced with as many changes as we are experiencing now, or expect to see within the next five

[13] Mattlage 1999; Odlyzko 1999.
[14] AUCC 1995.
[15] Dowling 1997; Halporn 1997.

to ten years.' Creating new models of digital 'self-publishing' is often seen as the way forward.[16]

The advent of the electronic journal has sparked all kinds of fantasies about the future of scientific communication. These fantasies concern mainly two issues. One is the relationship between academia and commercial publishers. The other is the use of properties that are specific to the *digital* nature of electronic journals for creating innovative modes of communication.

Academia and commercial publishers

The first issue is about the role of (commercial) academic publishers and the extent to which digitization might offer opportunities for publishing by the academic community itself, and hence its 'liberation' from control by what are perceived by some as capitalist, profit-seeking global companies. Kling and Callahan (2003), in a recent analysis of scientific electronic journals, summarize the views and expectations held by analysts of these developments, and relate them to problems concerning peer review and the economic crisis in scientific publishing. The rising cost of scientific publications, and the increasing problems surrounding copyright and control of intellectual property in the digital domain have acerbated the debate. Much of the discussion in this area is focused on creating new business models for scientific publishing, including various modes of 'self-publishing' by the academic community.[17]

The e-journal as an innovative format

From the very beginning, advocates of the electronic journal have stressed its innovative properties in addition to other aspects such as speed of production, lower cost and ease of distribution. Some expectations went even beyond traditional concepts of the scientific document as a discrete entity, and emphasized concepts of interrelated 'knowledge objects' and 'knowledge networks'.[18] The key innovative properties as commonly described in the literature can be summarized as:

- the use of multimedia (e.g. moving images and sound)
- enrichment of the communications format (e.g. through inclusion of datasets, embedded software etc.)
- new navigational modes, especially incorporating hyperlinks

[16]Tananbaum 2003. For additional discussions of the serials crisis see: Houghton 2001; Kaufman 1998; Mobley 1998.
[17]Mackenzie Owen 2002.
[18]Chien 1997; Hibbitts 1999; Mackenzie Owen 1987.

- increased user involvement and interaction
- new distribution formats (e.g. continuous addition of articles instead of serialization in volumes and issues), etc.

Over the past fifteen years a certain number of scientific journals have indeed utilized some of the opportunities offered by the digital format.[19] However, the impact of these journals has been masked by another development that in itself is highly significant: the move by the major academic publishing houses to publish their existing printed scientific journals in digital form over the Internet. This development has resulted in the creation of new and highly sophisticated mechanisms for retrieval, linking, access control, delivery and licensing. It is also most probably leading to significant changes in the practice of information use and scholarship, perhaps even with some impact on the choice of research topics and methodologies. Such effects might range from an increase in the volume of citations due to easier access to the scientific literature, to a shift towards research areas where digital resources are more easily available than in others. But in the majority of cases the journal article as a communicative form for reporting on research and for disseminating scientific knowledge does not seem to have been transformed by this mode of digitization: it remains a digital copy of the printed form.

Peek and Pomerantz (1998) found that academic publishers have embraced the Portable Document Format (PDF) to create a mirror image of the printed version. They also found that the PDF format was not used to innovate scientific publishing:

> 'However, the PDF format is not being used to its full potential. Adobe Acrobat allows objects to be embedded into PDF formatted documents. These objects can include those common on the World Wide Web: Pop-up lists, radio buttons, dynamic and interactive control objects that trigger sound and movie files, as well as the most familiar object on the Web the hypertext link. However, these objects are rarely to be seen in the PDF documents that we observed. Rather, these publishers are producing PDF files that entirely emulate the printed page, down to its static nature. ... The promise of electronic publishing is one of a fully interactive and hyperlinked environment. PDF, as it is currently being used with these publishers, reduces the benefits promised for electronic publishing'.[20]

Therefore, in order to understand the way digitization might lead to a transformation of scientific information media, we shall have to look elsewhere and seek out those media that have developed as digital forms without being bound to a printed counterpart. In spite of the claims cited above about the digital format's potential to 'revolutionize' scientific communication, it remains unclear if and in which way that potential has been realized. Several years after

[19] See chapter 5.
[20] Peek and Pomerantz 1998, p. 985-6.

the first e-journals it was noted by Kling and Covi (1995) that most of their articles might as well have been published in print journals: ‘few of [today’s e-journals] use special features of the electronic media to scholarly advantage. ... The articles ... do not make special use of their electronic formats’. And in a survey of the first years of scientific and technical e-journals Hitchcock et al., (1996) had to conclude that:

> ‘The immediate future of online journals is set to be dominated by electronic editions based on established paper journals and retaining the appearance of familiar paper layouts through Adobe Acrobat. The innovative features made possible by online publishing may therefore be obscured for a time. Their emergence on a large scale will depend on whether electronic editions can evolve to become electronic-only journals, or on the commitment of the scholarly community to demand these features through the development of new journals, producers and publishing structures’.[21]

In an early study carried for the JISC/eLib Programme in the United Kingdom, Eason et al., (1997) observed about the impact of electronic journals on scientific practice that:

> ‘... each discipline has very deep rooted reasons for the way it is constructed and the way scholarly activity is undertaken. Fundamental change will only come when the scholars are satisfied it will be in the interests of the discipline. The form of change will be determined from within the discipline rather than from outside sources ...’ [22]

So while there are many claims for a ‘revolutionary’ role of the electronic journal, empirical studies seem to indicate a more limited role predicated on conservative mechanisms and traditional cultures of practice within scientific communities.

4. Studying digitization

There are two ways to study the digitization of the scientific journal. One is to take the electronic journal as a fact of life and as a point of departure for research. This leads easily into the area of Technology Assessment (TA).[23] The underlying research question for such a study in terms of academic policy would be how the digital journal affects the practice of research, and in terms of traditional information science how the communication system should evolve to cope with the increasing availability of scientific information in full-text digital form. Such an approach would necessarily focus on researchers as *readers* and *users* of digital research information.

However, the main focus of this book is on a different issue, illustrating the second approach to the phenomenon of the electronic journal. Here we are not primarily concerned with the impact on research practice and information systems of the scientific journal as it migrates from print to digital, but with the

[21]Hitchcock et al., 1996.

[22]Eason et al., 1997, section 6.2.7.

[23]This is the approach taken by Nentwich (2003) in his study of ‘cyberscience’.

question to what extent scientific genres,[24] and in particular the research article are themselves transformed by the process of digitization. The basic question here is whether digitization leads to a form of communication that is fundamentally different. Does digitization only offer a new means for producing, archiving, accessing and navigating the traditional research article? Or does it lead to new forms of the research article that do not (and perhaps cannot) exist in non-digital, printed form? Fundamentally the question is whether the process of digitization has in fact resulted in a new, truly *digital* genre for scientific communication and has therefore changed the communicative practice of the researcher-as-author. Here, too, there are underlying questions related to academic policy and systems development. But whereas in the approach described above these questions relate to the way digital information resources are *used* (with subsequent effects, e.g. on the choice of research topics or citation behaviour), our approach has a different emphasis. Here the focus is on the effects of digitization on the scientific communicative practice, i.e. on the information resources *created* by researchers. In short: does digitization lead to new modes of reporting on research?

The two themes that inform this book are therefore (1) the nature and properties of the digital format as used by scientists and (2) the extent to which these properties affect the practice of (formal) scientific communication. The first theme addresses the properties of the digital as embodied in new publication formats within the scientific domain. For instance, there is a sense that digital versions of printed journals are not 'really' digital. 'Real' digital journals are thought to be journals that have no print equivalent, and that are therefore not bound to the constraints of the traditional printed journal.[25] But in what way are these 'real' digital journals then different? In other words: what can be defined as the singular properties of the digital? And to what extent are these properties actually embodied in scientific articles published in digital form?

This theme involves issues that are more sociological and behavioural in nature. If it is true that digitization – as is often asserted – has had a fundamental, even revolutionary impact on scientific communication, what would this amount to? Surely this would have to involve a fundamental change in the communicative practice that fully utilizes the properties and potentials of the digital format. Therefore, our approach will be – as described in more detail later in this chapter – first to define those properties and potentials, i.e. to theorize about what the impact of digitization on the scientific information medium might be. Then we address the more behavioural issues – the way in which researchers use the digital format – by analyzing existing electronic

[24] For the concept of *genre* in information science, see Orlikowski and Yates 1998; Hjørland 2002a; Bazerman 1988.

[25] Cf Kling and McKim 1999.

media in terms of our theoretical considerations. This will allow us to draw conclusions about the extent to which digitization has in fact transformed the communicative practice of researchers.

Again, within the field of scientific communication, there is a sense of 'revolution', of drastic change and a radical departure from the traditional practice. Nentwich (2003, p. 3) points to the frequent comparisons with the effects of the invention of the printing press by Gutenberg, referring to Harnad's well-known article on the 'Post-Gutenberg galaxy'.[26] At the same time, there is a feeling that nothing has really changed at all, or at least that there is a reluctance if not opposition to adopting change within the scientific community. This leads us to the question whether, or to what extent and in which way the practice of scientific communication has been 'digitized'. Within the scope of this book this will be operationalized as *the extent to which researchers utilize new, digital means for reporting on their research outcomes through the peer reviewed article.*

5. The information science context

The issues that we shall examine in this study are usually assumed to belong to the field of information science. In view of the many, diverse and sometimes conflicting definitions of this domain, it is necessary to specify more precisely the position taken here.

Marcia Bates (1999) has described information science as 'the study of the gathering, organizing, storing, retrieving and dissemination of information'. She describes as the domain of information science 'the universe of recorded information that is selected and retained for later access' and 'produced by human agency', adding that 'we find ourselves primarily concerned with the form and organization of information, its underlying structure, and only secondarily with its content.' Bates goes on to identify three 'big questions' that should be addressed by information science:

- the physical question: the properties of recorded information and the laws governing its universe
- the social question: the ways in which people relate to, seek and use information
- the design question: how to achieve rapid and effective access to recorded information

[26] Harnad 1991. Other discussions of the 'Gutenberg revolution' in the context of electronic media include: Birkerts 1994; Füssel 2001; Giles 1996; Hammes 2001; Siler 2000. We shall return to this issue in the next chapter.

The subject matter of this book is related to the first two of these questions. We shall seek to analyze the properties that characterize (a specific form of) *digital* information. We shall also investigate the extent to which people (in our case: researchers) use these characteristics in the information they create and use. But there is a difference in approach that relates to the conception of the 'universe of recorded information' underlying the outline of information science given by Bates and others. Our main concern is with information *in action*, i.e. as performing a role within scientific communication. A research universe limited to the domain of information as an isolated entity is therefore too constrained. It focuses, as does Vickery's conception of information science described in the next section, on information as a static object (and the functions performed on it) that is a valid object of research on its own, without regard to the context in which it operates. Such an approach neglects the role of information as (component of a) process, without which it loses its significance.

The difference in approach can be illustrated as follows. In 'conventional' information science the concept of an electronic journal would be taken as the point of departure, using actual electronic journals as the objects of analysis. These would then be analyzed as to their properties, the way in which they are used, the systems required for their management, etc. In this book, we postulateregard the electronic journal as a construction that results from the way a social entity (researchers) utilize technical possibilities (the outcome of the process of digitization) when they disseminate their knowledge in the form of recorded information. The point of departure is therefore a more abstract conception of 'the digital' in the context of a scientific communication practice.

Harold Borko describes information science as:

> 'that discipline that investigates the properties and behavior of information, the forces governing the flow of information, and the means of processing information for optimum accessibility and usability. It is concerned with that body of knowledge relating to the origination, collection, organization, storage, retrieval, interpretation, transmission, transformation, and utilization of information'.[27]

He also notes that 'it has both a pure science component, which inquires into the subject without regard to its application, and an applied science component, which develops services and products'. This leads to the question what the relationship is between the outcomes of the 'pure science component' and the 'applied science component'. Hjørland, in spite of his philosophical and metatheoretical inclinations, seems to have an applied science approach in mind when writing that 'information science is concerned with research which might help improve the design of information systems and services'.[28] He goes on to say:

[27] Borko 1968, p. 3.
[28] Hjørland 1998, p. 618.

> 'However, the communication systems of science (and other social systems) are much older than the computer age, and have, through centuries, developed important characteristics such as source criticism, principles of rhetoric, standards for publishing, and so on. All this represents production, dissemination, and use of information, which is the declared object of research in information science. The understanding of this social system is a pre-condition for establishing computer-based systems to make the system more efficient. If this kind of knowledge is lacking, systems design might be a mistake. ... One important perspective is therefore represented by those investigations, which analyse the historical developments of information systems as adaptations to specific communicative needs. Such studies can be published under many labels, one of which is social constructionism'.[29]

This extensive quotation points to the position of this book within the broader domain of information science. To the extent that scientific communication is open to design and engineering, and more specifically can be changed and improved by the application of ICT, some understanding of the social system of scientific communication and its historical development is required. The digitization of scientific communication can be seen as a (recent) historical process in which a specific relationship between the adoption of digital technology and the use of information genres has developed. It is this relationship that places our study in the 'pure science' component of information science as described by Borko, but with the expressed intention of providing a background for a social constructivist approach to scientific communication. In contrast with much information science research, however, we shall emphasize the evolving nature of scientific information itself rather than the functional processes (such as origination, collection, organization, storage, retrieval, interpretation, transmission etc.) that operate *on* the information.

6. Theory in information science

Michel Foucault, in *The archaeology of knowledge*,[30] describes the concept of *discursive formations*: 'whenever, between objects, types of statement, concepts, or thematic choices, one can define a regularity (an order ...), we will say, for the sake of convenience, that we are dealing with a *discursive formation*'. Radford (2003), following Wiegand (1999), notes that information science is 'trapped in its own discursive formations.' The idea here is that information science is practiced within a fixed framework of concepts, vocabularies, theories – with little regard for the fact that what is now accepted has not always been, and will not always be the case. However, information science is also often criticized for its deficiency in terms of theoretical underpinning,[31] suggesting that the problem is more a *lack* of a coherent discursive formation. Pettigrew and McKechnie found that, whereas there appears to be

[29] Hjørland 1998, p. 619.
[30] Foucault 2002, p. 41.
[31] Hjørland 1998; Templeton 1994; McGrath et al., 2002; Warner 2001b.

some increase in the use of theory in articles published in information science journals, there are also clear discrepancies in the way researchers working in different subfields define the concept of 'theory' itself.[32]

The issue of theory in information science was put up for discussionby Vickery in 1997 in a well-known article on 'metatheory and information science'.[33] Metatheory is described by Vickery as the analysis (or rather the explication) of the concepts and assumptions underlying a knowledge domain. His analysis identifies a set of constructs that are thought to be central to the domain of information science and about which a list of 35 presuppositions centers that together form his proposed metatheory.

Table 1.1. Vickery's information science constructs

- information as knowledge-containing messages
- structure and representation of knowledge
- relevance and contribution to the process of informing
- message designations (i.e. content indicators)
- mental and system models
- language
- information want
- query expansion and modification

Table 1.1 lists Vickery's constructs, illustrating a view of the information science field that is based on the conventional information retrieval approach in which the research domain is limited to a process of question answering of which the outcome is a set of documents that satisfies an information want expressed as a query. In general terms this approach views a set of documents D that contains a subset D^i satisfying a query Q^i. Information science is then seen as having as its goal to satisfy the function $Q^i \Rightarrow D^i \subset D$ through the application of its various constructs. This is a conception of theory that lacks any sociological grounding by abstracting from human and institutional actors in the information domain.

[32] Pettigrew and McKechnie 2001; McKechnie and Pettigrew 2002.
[33] Vickery 1997.

Vickery's listing of the presuppositions of information science – to be regarded as a first attempt – was found to be too general by Hjørland (1998), who has proposed a more philosophical approach notably based on an historical and epistemological analysis. Hjørland's contribution to the metatheoretical debate illustrates a recent shift from the traditional grounding of information science in a positivist systems theory towards an approach informed by post-positivist and other philosophical discourses.[34]

Before explicating the theoretical grounding of this book, we need to have some understanding of what we regard as a theory in the context of information science. Although providing a comprehensive definition of the concept of a 'theory' has been described by Ziman as 'an invigorating but fruitless walkabout in metaphysics',[35] some idea of what we are talking about is clearly necessary. Here we follow Smiraglia who defines theory as 'a system of testable statements derived from research':

> 'Theory is derived from the controlled observation of phenomena, whether this has taken place in the positivist empirical paradigm or in the qualitative paradigm. Theory is the basis for research, serving to supply hypotheses for empirical research, and to confirm observations in qualitative research. The power of theory is its explanatory capability. We can use theory to analyze, predict and manipulate phenomena.'[36]

Glaser and Strauss describe three roles that are performed by theory: to provide control over situations, to provide a perspective of behavior and to provide a contextual framework for research.[37] It is the latter role that we are seeking here. Glazier and Grover describe theories as 'generalizations that seek to explain relationships among phenomena', a 'multiple-level component of the research process, comprising a range of generalizations that move beyond a descriptive level to a more explanatory level'.[38] In other words, theories arise from the observation and analysis of phenomena to constitute mental models that help us to understand and predict how the world operates. Michael Buckland also describes theory as a mental construct, viz a construct that should match our perception of the observed phenomena as well as possible.[39] A first step in specifying a theory should therefore be a delineation of the phenomena that form the theory's domain of analysis. An extended theory will then provide a language and contextual framework, including a typology and some explanation of the *dynamics* of the phenomena, i.e. of changes over time and/or space.

[34]See for instance Budd 2001; Day 2001; Hjørland 2002b; Raber and Budd 2003; Trosow 2001; Warner 2001a.
[35]Ziman 2001, p. 117.
[36]Smiraglia 2002, p. 331.
[37]Glaser and Strauss 1967.
[38]Glazier and Grover 2002.
[39]Buckland 1991a, p. 19.

The phenomena that we shall study here are, as explained above, the process of digitization, the research article as it is transformed through digitization, and the communicative practice of researchers as reflected in their use of digital means in the context of the scientific article. Our approach is to observe the phenomenon of digitization to develop a theory that specifies the properties of the digital research article. This theory will provide us with a hypothesis about what to expect if researchers – in their role as authors – fully use these properties. We then observe the actual use of these properties in the digital research literature, which will allow us to draw at least tentative conclusions about the communicative practice. At a broader level, these conclusions constitute a theory about the academic communicative practice that could be tested – by other methods – in a subsequent study.

7. Methodology

The perceived lack of a theoretical grounding in information science may be partly due to an insufficient understanding of the methodological issues involved in the relationship between theory and practice. The idea that information science is an applied science that ultimately aims to improve the dissemination of information through the development of retrieval systems fails to acknowledge the need for a theoretical grounding and a proper methodological process that leads us from an understanding of basic principles to practical applications. To quote Chatman:

> 'As researchers who wish to develop a theory, we must identify problems central to our field. ... once these problems have been identified, we might be led to the formulation of conceptual issues that underly these problems. This strategy is commonly referred to as the inductive method. ... it would appear that we are currently focused on the application of conceptual frameworks rather than on the generation of specific theories'.[40]

Information science research is often classified as either *quantitative* or *qualitative*. But as Tom Wilson (2002) has rightly pointed out, quantitative methods can play an important role in qualitative research, and vice versa. A more useful distinction is that between positivist and humanistic approaches. Wilson describes the positivist view as one 'in which social facts can be known with certainty and in which the laws of cause and effect can be discovered and applied'. The humanistic approach is informed by the idea that social reality is constructed though meaningful action, and that meaning therefore is always grounded in a social context.

Wilson goes on to say that the observation of phenomena as the foundation of research can then be classified as either *direct* (observing the phenomena themselves) or *indirect* (studying accounts or records of the phenomena). Both

[40]Chatman 1996, p. 193.

direct and indirect observation can be further classified as having a structure that is either *imposed* (by the researcher, e.g. by means of a questionnaire) or *emergent* (i.e. emerges from the data at hand). When research is exploratory and aimed at developing theory, an emergent approach to structure is to be expected.

Since this book is aimed at clarifying the meaning and significance of digitization in the context of the scientific communication practice, we adopt a humanistic rather than a positivist approach in Wilson's terms described above. This approach can also be described as phenomenological and interpretative (i.e. based on observation of what actually occurs in social reality), as well as qualitative and exploratory. The nature of our subject matter – comprising a theoretical construct ('digitization'), an artefact (the scientific article) and a social practice (scientific communication) necessarily implies a certain degree of methodological pluralism.[41] Understanding the fundamental nature of digitization in relation to scientific media requires an indirect, inductive and necessarily exploratory approach. Analyzing the extent to which digitization is reflected in social practice (i.e. in scientific publishing) requires a more direct and deductive approach. In this respect we follow the suggestion by Kling and Lamb (1996) that the prevalent utopian, technological deterministic view of scientific communication needs to be counterbalanced by 'a different investigative strategy and genre of reporting analytical insights ... based on examining existing electronic formats as they are actually used in real social settings'.

Our approach is predicated on what Steve Woolgar has described as 'analytic scepticism': an approach that does not take the claims of technology for granted, but employs a combination of theoretical questioning and detailed empirical study to achieve a balanced and realistic assessment of the process and impact of technological development.[42] This book questions the extent to which digitization has impacted on the substance of scientific communication. It is informed by the argument that if this is the case, examples will have to be found in articles published in 'digital-only' journals that have been created specifically to utilize and benefit from the digital format (i.e. in contrast with e-journals that are digital copies of print journals).

8. Level of analysis

An important issue that has to be addressed in a study of this kind is the level of analysis. Tom Wilson has argued that the methodology of research in information science needs to be related to the integrative level[43] of the

[41] Wildemuth 1993; Tashakkori and Teddlie 1998.
[42] Woolgar 1999.
[43] Foskett 1978; Gnoli 2003; Korpela et al., 2002.

analysis.[44] For instance, in taking an all-encompassing view under the title of 'cyberscience', Michael Nentwich has employed methods of technology assessment (unfortunately with insufficient empirical underpinning) to create a broad canvas of the impact of information and communication technologies on the practice of science, with an emphasis on scientific communication.[45] The level of analysis adopted in this book, however, is far more detailed. It is aimed at a deepening of our understanding of a core phenomenon of scientific communication: the article as a carrier of knowledge acquired through research. In fact our aim is even more specific: *to arrive at an understanding of the way digitization transforms the research article as a scientific genre.* Our approach is therefore more concerned with a micro-level of analysis than with a macro-level.

However, the scientific article is not an isolated entity but a component of a social system that provides the context for scientific communication. This context is described in Fig. 1.3.

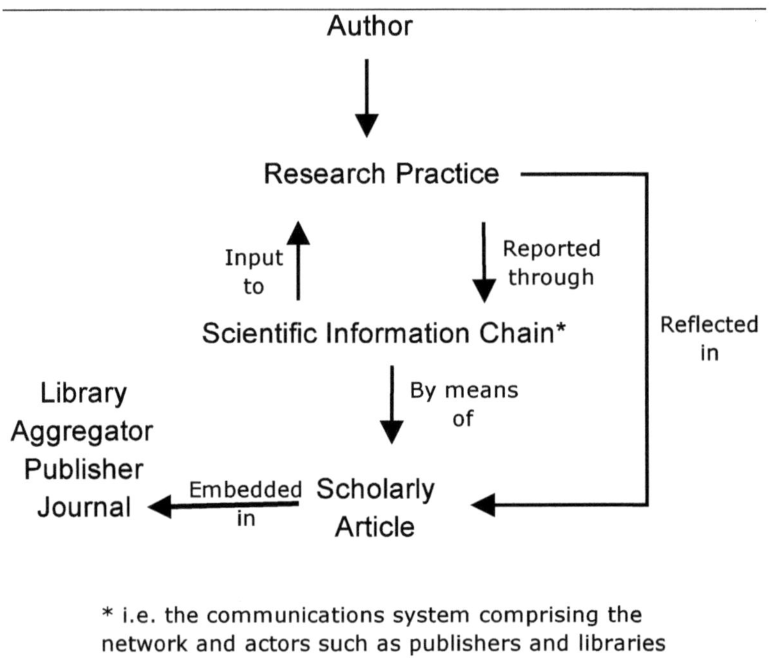

Figure 1.3. Research domain

[44]Wilson 2002.
[45]Nentwich 2003.

Table 1.2. Outline of the book

Chapter	Topic
1	Introduction: subject, theory, methodology, level of analysis
2	Scientific communication: historical perspective
3	The scientific communication system: analytical perspective
4	The concept of digitization: a metatheory of digitality
5	Analytical study: the digital research article
6	Summary and conclusions

Our primary object of analysis is the scientific article as the representational output of research by an author (defined as an individual researcher or a research group). The context for scientific communication is therefore defined by the boundary constructs *author* and *research article.* The scientific article is an information resource embedded in a multi-layered system that we shall analyze in more detail in chapter 3. In brief, the article forms part of a container construct, the scientific journal. Journals are often grouped by publisher, whereas journals from various publishers are grouped in access points provided by 'aggregators'.[46] These groupings are relevant because properties of the individual article are sometimes derived from properties of the higher, aggregated levels. In other words, what the reader perceives as a property of the article he or she is reading, may be a property of the individual article or it may be a collective property inherited from a higher level and available to all articles within that level. Further context is provided by what we refer to as the 'scientific information chain', described in chapter 3. This can be regarded as the systems layer through which research information is transferred from the researcher to publicly available information resources and back. The information chain comprises both the network and various intermediary actors such as publishers and libraries. A final level is that of the research practice, defined as the aggregate of individual research activities of authors participating in the scientific information system.

9. Outline of the book

In chapter 2 we investigate the development of the scientific communication system as the context in which the scientific document functions. Attention is given to the functions and properties of the system, the nature of the scientific

[46]See page 74. Examples are companies such as EBSCO, Ingenta or Emerald.

document, and the actors involved. We also trace the development in time as a function of increasing digitization (a theme we shall return to in chapter 4).

In chapter 3 we analyze the development of scientific communication and the impact of ICT in a more structural way. To this end we use a number of model-based perspectives that help to describe various characteristics of the scientific communication system. We use this model-based approach to develop a model of innovation of the scientific journal. Finally, we show how the perception of the relationship between scientific communication and ICT has shifted over time.

To conclude the theoretical part of the book, we turn to the concept of digitization in chapter 4. In this chapter we develop a set of concepts as a metatheoretical 'language' for describing and understanding digitization in the context of scientific communication. From this we derive a set of hypotheses about what the digital research article might be if it were fully to utilize its digital properties. Our aim is then to extend our analysis into the realm of social reality, by investigating to what extent these digital properties actually 'materialize' in the practice of scientific communication. This is done in chapter 5 by developing an analytical framework from the findings of the preceding chapters. This framework is then applied to a study of selected digital-only, peer-reviewed scientific journals and their content.

Finally, in chapter 6 we summarize our findings and draw some conclusions with regard to the impact of digitization on scientific communication, based on a number of theoretical perspectives on the impact of technology on society in general, and scientific communication in particular.

Chapter 2

THE DEVELOPMENT OF SCIENTIFIC COMMUNICATION

In their well-known book on the emerging network society, Michael and Ronda Hauben wrote in 1997:

> 'A revolution in human communications is happening. [...] Comparing the emergence of the printing press to the emergence of the global computer network will reveal some of the fascinating parallels which demonstrate how the Net is continuing the important social revolution that the printing press had begun. [..] Just as the printing press essentially replaced the hand-copying of books in the Renaissance, people using computer networks are essentially creating a new method of production and distribution of creative and intellectual written works today'.[1]

As mentioned in our introductory chapter, many accounts of current developments in communication describe the process of digitization and the use of networks as a 'revolution' – often referring back to the invention of the printing press in the 15th century and the ensuing so-called 'Gutenberg revolution'.[2] The Hauben's 'Netizen' is but one example of this mode of thinking. Scientists themselves are less outspoken about the current state of affairs, but nevertheless sometimes express faith in the way technology might change things. For instance, Steven Bachrach, professor of chemistry at Trinity University writes:

> 'Yet, while all of these changes have radically altered our society, the primary means by which scientists communicate with each other has remained frozen in time, unchanged in well over 100 years. We scientists still produce the written article, published in the specialized scientific journal, that appears as ink applied to paper. [...] And yet, even though the use of color graphics,

[1] Hauben and Hauben 1997, ch. 16 (http://www.columbia.edu/~rh120/ch106.x16).

[2] The historical analogy in utopian technological accounts is not confined to printing, nor to the occasional eccentric, as is shown by the ideas of the European Commission about the perceived inevitability, manageability and benefits of the information society. See for instance Jean-Claude Thébault's foreword to a study on democracy and the information society in Europe (Hubert and Caremier 2000, p. x). Here, information technology is seen as instrumental in the development of citizenship and democracy, depicted as a 'revival of an old idea', a 'virtual forum' comparable to the Republic of the Humanists in the 16th century.

animations, sound, and extremely large data sets are now routine and essential components of the scientific method and process, all of these are omitted when the time comes to distribute the knowledge among our colleagues. Since we write journal articles destined to appear in print on paper, we can't include a movie or sound. Color images are still (for most journals) too expensive to print. Large data sets consume a significant number of the limited pages available in journals and are at best relegated to supplements, if not left out entirely. [...] The time is ripe for a dramatic, profound shift in how scientists should (and will) communicate in the near future'.[3]

According to Bachrach and many others we are to expect nothing less than a radical departure from traditional practices, and a communications system transformed by information technology. But how realistic are such expectations? Can we sensibly compare the current time frame of less than two decades to the period described by Eisenstein in her study of communications and cultural transformations in early-modern Europe?[4] Is it a plausible idea that a new technology will drastically change a communicative practice that has evolved over the centuries? These are the questions that we shall address in this chapter.

The structure of this chapter is as follows. We begin with the origins of scientific communication until the Scientific Revolution of the 16th and 17th centuries, followed by a discussion of the significance for science of printing as an early example of a 'communications technology'. We then describe the role of the scientific societies and the development of the scientific journal as the primary form for formal scientific communication. This is followed by an analysis of the current development towards digitization of the scientific journal and of the evolution of the scientific article as a communicative genre.

1. The historical perspective

The desire to add to one's stock of knowledge from what others already know, the quest for information or 'intelligence', is a characteristic of scholars and scientists of all times. In 17th century Cambridge it was already observed that 'scholars are so greedy after news ... that they neglect all for it'.[5] The most basic form of knowledge transfer is by word-of-mouth: information brought by a messenger from one location to another; using the written word is a later innovation. The earliest form of (informal) scientific communication was therefore the face-to-face transfer of knowledge from a master to his pupils. These pupils then went out into the world to share their acquired knowledge with others and thus spread their master's knowledge even further. But we do know that already at a very early stage – at least as early as the 7th century BC – the *written* word was used for creating documentary records that contributed to the advance of knowledge throughout the civilized world.[6]

[3] Bachrach 2001.
[4] Eisenstein 1980.
[5] *Publick Occurrences or news from both city and country*, July 7, 1679, quoted in Stephens, 1989, p. 13.
[6] Vickery 1997, p. 1-4.

The initial use of the handwritten scholarly document and of its printed successors from the mid 15th century onwards was mostly confined to the consolidation and transmission of existing, authorized knowledge, as was the common practice in the early universities. Scientific communication as we understand it nowadays – the exchange and discussion of *new* knowledge derived from the author's own research and based on observation and verifiable facts – depended initially on interpersonal exchanges between individual scholars, based on correspondence and travel. It was not until the mid 17th century – and initiated by the early scientific societies rather than the universities – that the printing press was employed for a major innovation in scientific communication: the exchange of ideas and results of empirical research, as well as their public debating, through the medium of a periodical, subscription-based publication. Since these early beginnings the diffusion of scientific knowledge has developed into a global business based on the newest digital technology without which science itself cannot function. To understand the nature of these developments we must go back to the origins of modern science and the so-called Scientific Revolution.

1.1 The Scientific Revolution

Modern science has developed through what is commonly known as the *Scientific Revolution* of the 17th century,[7] described by Kearney as 'all the imaginative achievements associated with the names of Copernicus, Galileo and Newton, ... a revolution ... in the way in which men regarded the universe'.[8]

A diffuse concept,[9] the historical boundaries of which are anything but clear, the Scientific Revolution denotes the transition from the scholarship of the Middle Ages and the Renaissance to the experimental 'natural' sciences of the 16th and 17th centuries. The scholarly practice before the Scientific Revolution was based on the collection, analysis and consolidation of received, pre-existing knowledge, including the translation of Greek, Hebrew and Arabic texts into Latin. Handwritten scholarly documents and their printed successors from the 15th century onwards were the vehicles for recording and transmitting this type of knowledge. Stephen Jay Gould illustrates this practice in his description of Gesner's *Historia animalium* of 1551 as:

> '... not a scientific encyclopedia in the modern sense of presenting factual information about natural objects, but rather a Renaissance compendium of everything ever said or reported by human observers or moralists about animals and their meanings, with emphasis on the classical authors of

[7] Our sketch of the Scientific Revolution is based mainly on Burke, 2000; Hatch, 2002; Hooker, 1996; Shapin, 1996; Westfall, 1986, p. 38-44.
[8] Kearney 1966, introduction, p. xi.
[9] Shapin introduces his well-known study with the statement 'there was no such thing as the Scientific Revolution, and this is a book about it'. (Shapin, p. 1).

> Greece and Rome (seen by the Renaissance as the embodiment of obtainable wisdom in its highest form), and with factual truth and falsity as, at best, a minor criterion for emphasis'.[10]

The Scientific Revolution was predicated on the transition from this type of compendium-based scholarship to a practice based on exploration, discovery and the creation of new knowledge. Leydesdorff (2001) has described this transition as a switch from a normative belief system to a system based on expectations in which 'truth can be investigated, and thus the search for truth can function as a code which guides the communication'. As a consequence, the sciences 'have been socially constructed as discursive systems of rationalized expectations'.

It is unquestionable that this was a transition of major consequence for the modern world. In that sense the term 'revolution' is entirely justified. But a *fast* revolution it certainly was not, spanning a century and a half – beginning with Nicolaus Copernicus early in the 16th century, followed by scholars such as Bacon, Kepler, Gilbert, Galileo, Descartes and Huyghens amongst many others, and culminating in the work of Boyle, Hook, Halley and especially Newton.

It has been common practice since the 17th century to think about science as an 'open' communicative system based on the widest possible diffusion of ideas and research findings, allowing for their unrestricted scrutiny, criticism and debate. This implies some mechanism that ensures the proliferation and exchange of ideas, and the universal availability of research outcomes irrespective of time and place. Such a mechanism has to be seen as an organized system for scientific communication that performs specific functions (e.g. distribution, access, preservation), and that also operates as a social system in that whoever is excluded from the communicative system cannot act as a member of the scientific community.

It is tempting to equate the beginning of this system with the long Scientific Revolution. But in fact the early universities already constituted a communication system that shared many characteristics with its modern counterpart. The universities served as 'clearinghouses' for the exchange of information,[11] both orally and in written form. They also contributed to the social contextualization of scholarship through frequent exchanges between academia and the societies in which they operated. In addition, and in contrast with the earlier monastic centres of scholarship, they were characterized by a degree of interdisciplinarity that would also be a characteristic of the Scientific Revolution before specialization in the 19th and 20th centuries destroyed the unity of the new intellectual project.

[10]Gould 2004, p. 2.

[11]The concept of a *clearinghouse* is discussed in chapter 3.

1.2 The early impact of printing

If we regard the printing press as an early example of an 'information and communication technology', what was its significance for the development of science? Initially, before the invention of moveable type, scholarly manuscript books used for the distribution of knowledge (i.e. as opposed to archival manuscript sources) were often produced by *stationarii*[12] on the basis of *exemplars:* manuscripts authorized by the university where the stationarius had set up his trade. The invention of printing in the middle of the 15th century did not immediately change this practice. In the beginning, printing functioned more or less as a 'mechanisation' of the handwriting practice. Although the *quantity* of reproduction increased dramatically, the substance and communicative *function* of printed works were not much different from those of the manuscript book.

The initial practice of the early scholarly press therefore was to follow the *established* practice on a larger scale, making existing, authorized knowledge available to a wider public than was possible before the invention of printing. It is of course true that the invention of printing – predating the high point of the Scientific Revolution by two centuries – created opportunities for the dissemination of information in academic circles throughout Europe. However, universities did not readily regard in-house printing as something 'of their business'. Although the first book was printed in Oxford in 1478, the university did not acquire a printing privilege until 1586.[13] Printers, on the other hand, did not regard universities as financially attractive institutions.[14] Publishing was at its best based on a direct relationship between scholar and printer.[15]

The impact of printing on academia was initially more quantitative than qualitative, resulting in consolidation rather than in innovation. In this sense the concept of a 'Gutenberg revolution' is misleading, at least within the context of scholarly communication. Apart from creating a wider audience for the existing scholarly practice,[16] the printing press did not immediately lead to a *new* practice. That was to come much later. That is not to deny, of course, that in the *longer term* printing did act as a change agent (amongst many other factors) in the development of the Scientific Revolution. For instance, Elisabeth Eisenstein points to the fact that the enhanced availability of texts (i.e. the quantitative effect of printing) resulted in a heightened awareness of the

[12]The *stationarius* performed the various functions associated with books: printing, publishing, selling, and lending. He would also act as a stationer, selling writing materials to scholars and students. Stationarii set up business within the university.

[13]OUP 2004.

[14]Vickery 2000, p. 60.

[15]See also Hunter 2001.

[16]Including a non-scholarly readership of literate laymen outside the realm of academia.

inconsistencies in classical knowledge with, as a consequence, a controversy between 'literalists' and 'modernists'.[17]

The lack of an immediate innovative and 'revolutionary' effect of printing in this domain is explained by William Eamon in an interesting way that is relevant to our analysis.[18] Eamon argues that printing eventually enabled the Scientific Revolution by making a far greater amount of *input* data available to researchers, i.e. by facilitating access to a much larger array of facts, opinions, methods etc. (often from non-scholarly, empirical sources) than was possible with non-printed sources.[19] In other words, the important function of the printing press was not that it facilitated the dissemination of research output, but that it enhanced access to data sources in the input phase of research.[20] Since it required a considerable amount of time to create a significant body of data in printed form, it is understandable that the print 'revolution' in the domain of science was a relatively slow process.[21]

1.3 The scientific societies

Although the invention of the printing press did not *immediately* result in a new scientific practice – and in many respects it reinforced the prevailing mode of scholarship – it is nevertheless clear that the printed book also contributed to the wider dissemination and further development of the new 'Baconian sciences' as they developed alongside the traditional scholarship practiced in the universities. Pyenson and Sheets-Pyenson describe the advance in communications technology brought about by the invention of movable type in the middle of the 15th century as one of the decisive enabling factors for the Scientific Revolution.[22] It is interesting to note that their argument derives not only from the role of print as a new, fast and pervasive mode of distribution, but also from the intrinsic characteristics of the print format itself. The new science demanded a high degree of accuracy, both between copies and in the text, numbers, diagrams and illustrations. It also required flexibility in incorporating new findings in revisions of the text. These were provided by the

[17]Eisenstein 1980, p. 523.

[18]Eamon 1994, Introduction.

[19]See also Eisenstein 1980, p. 520 ff. Eisenstein also points to the shift from words to facts, or from textual information to data (tables, charts, diagrams, maps etc) that was facilitated by printing, i.a. because of the increased accuracy or reproduction.

[20]See the discussion of research phases in chapter 7 on page 66.

[21]There is a possible analogy here between the role of print and that of digitization. Digitization enhances publication and dissemination of research results (just as print did), and it is tempting to seek its importance in this area. But the impact of digitization on the *practice* of science may be more due to the way it makes data sources available for scientific research. The extent to which digitization enhances access to data sources as *input* to scientific research may have a greater impact on the practice of science than its deployment for the dissemination of research *output*. Especially in the humanities, where the digitization of information sources is expensive and time-consuming, this may be an important issue.

[22]Pyenson and Sheets-Pyenson 1999, p. 215.

printed book, together with an enhanced openness through the wider public availability of printed works. It was through these printed books that the ideas of Copernicus, Bacon, Kepler, Galileo, Descartes and Newton – to mention only the most influential authors – were disseminated throughout Europe.

However, in spite of these innovative properties, the printed book ultimately did not prove adequate for coping with other characteristics of the new science, such as its speedy development, proliferation of activities and international character. The need for an alternative becomes apparent from the beginning of the 17th century onwards, initially resulting in a practice of scholarly communication that was based on personal networks grounded in extensive traveling and face-to-face meetings, lectures, and – with the development of postal services – the exchange of letters.[23]

Letters became an important communications medium in circles of the new Scientific Revolution in two ways. Firstly, they were used to write up technical and observational information on scientific issues and experiments – in fact as a precursor of the scientific article. Secondly, they were sent out in multiple copies to and distributed amongst larger numbers of scientists and interested amateurs, resembling a modern 'mailing list' and informational network. Some scientists specialized in compiling such 'learned letters', taking on the role of 'trafickers in intelligence'.

Face-to-face communication eventually became institutionalized on a scale that transgressed the boundaries of the individual through the establishment of 'scientific societies' such as the *Royal Society* in London.[24] Here the dissemination of scientific and technical letters also became institutionalized, as officials of the societies collected information that was sent out to their members.

The Royal Society of London was founded in 1660[25] by a group of men (including Wren and Boyle) who had already been meeting since the mid-1640's to discuss matters of philosophy.[26] The *Académie des Sciences* in Paris came just a few years later, as a group of scholars gathering fortnightly in the king's library in the rue Vivienne, and formally founded by Colbert in 1666.[27]

[23]Kronick 2001.

[24]Hartley 1960; Hunter 1989, 1994. See also the Scholarly Societies Project (http://www.lib.uwaterloo.ca/society/).

[25]Chartered by Charles II in 1662 as the *Royal Society for the Improvement of Natural Knowledge.*

[26]Boyle referred in his letters to the early meetings as 'our invisible college or the philosophical college'. For an interesting contemporary account see Wallis 1700, reprinted in Colby 1920, p. 196-199. See also http://www-gap.dcs.st-and.ac.uk/~history/Mathematicians/Wallis.html.

[27]The *Académie* remained a rather informal institution for over thirty years, until it obtained its first statute (as the *Académie royale des Sciences*) from Louis XIV in 1699. It membership was limited to 70 members and was involved both in publishing and in advising the government. After being suppressed by the Convention in 1793, its was replaced by the *Institut national des sciences et des arts* in 1795, with 144 members. The *Institut* was in fact an aggregate of the former scientific, literary and artistic academies (cf. Brown 1967).

This was just the beginning of a movement towards the establishment of 'broad church' and often national institutions that brought together the scientists of the enlightenment: the Accademia dei Dissonanti di Modena in 1683, the St. Petersburg Academy of Sciences in 1725, the Royal Swedish Academy in 1739, the Royal Society of Edinburgh in 1783, the Royal Irish Academy in 1785, the Koninklijk Instituut van Wetenschappen, Letterkunde en Schoone Kunsten in Amsterdam as late as 1808.[28]

The thesis put forward by Ornstein (1913) that the 'scientific societies' were created as a reaction to the hostile position of conservative universities towards new modes of scientific thinking is nowadays regarded as untenable.[29] Nevertheless there was an apparent need for a new type of institution that would bring together people both from within the university and outside, from various fields of science, and with a consciously expansive, outward-looking approach to the organization of scientific activity, as exemplified in endeavors ranging from novel forms of publication to scientific expeditions in the modern sense.

As channels for scholarly information, the scientific societies had a number of characteristics that came to be of importance for the development of the scientific journal:

- *Sharing*: the societies' role was that of an open forum for the exchange of ideas,[30] precisely because the new scientific method was based on critical scrutiny and validation of observations of the natural world. This was a crucial development in view of the traditional secrecy that scientists previously had maintained, a secrecy often necessary to prevent ideas being stolen by others.

- *Rapid diffusion* through the use of printing, periodical publication and the mail system.

- *Innovation:* the societies were the focal point for the creation and dissemination of new knowledge.

- *Reputational control:* the societies served a role in the legitimization of scientific claims, acting as a public registry of intellectual property (the Royal Society recorded on which date it received information from authors).[31]

[28]The scientific societies and their history are amply documented in the Scholarly Societies Project (http://www.scholarly-societies.org/).

[29]Burke 2000. On the presumed opposition of science and the humanities, see: Gould 2004.

[30]Pyenson and Sheets-Pyenson 1999, p. 74ff. This applies especially to the Royal Society and its Philosophical Transactions that was an exponent of the scientific method and the exposition of new knowledge. The Journal des Sçavans was a more 'journalistic' product that presented scholarly news in a somewhat gossipy form. Guédon (2001) describes this as a distinction between 'originality' and mere 'novelty'.

[31]Guédon 2001.

Both the open discussion (as a precursor of the peer review system) and the co-optation of members served to enhance the status of membership.[32]

- *Interdisciplinarity:* the societies encompassed the broad field of science, and ideas and methods were freely exchanged across what we now perceive as distinct disciplines. This situation continued until the explosion of specialization and diversification in the 19th century.[33]

- *Contextualization:* there was an interest in scientific application, and membership was also open to 'amateurs' working in applied fields. For instance, the Philosophical Transactions published reports on William Petty's double-hulled vessel, Robert Holme's use of Huyghen's clocks on the Atlantic voyage, and articles on chemistry, mining, agriculture, etc.[34]

- *Archiving:* The Royal Society and other scientific societies created an archive of their transactions, effectively documenting the progress of science and allowing reference to be made to specific observations and findings.

1.4 The scientific journal

The birth of the new (and Newtonian) science, the development of scientific communicative networks throughout Europe, and the establishment of the scientific societies form the context in which a new medium was created: the scientific journal. In January of the year 1665 the first issue of the *Journal des Sçavans* was published in Paris by one Denis (or Denys) de Sallo.[35] Its contents consisted mainly of book reviews, covering publications printed throughout Europe. In addition it contained scholarly discussions and reports on new scientific developments and conference reports. In the same year 1665 Henry Oldenburg, secretary of the Royal Society in London, set up (and financed) the *Philosophical Transactions* that eventually was to act as the vehicle for Isaac

[32]After the Royal Society took over institutional financial responsibility for the *Philosophical Transactions* in 1753, it established a committee for reviewing papers to be published, and adopted a regulation allowing the committee to ask the opinion of 'any other member of the Society who are knowing and well skilled in that particular branch of science that shall be the subject matter of [the] paper' (Kronick 1991, p. 5). This can be regarded as the beginning of the formal peer review system.

[33]Mono-disciplinary scientific journals were not introduced before the second half of the 18th century. See footnote 39.

[34]This was, however, more so in England than on the continent. The Philosophical Transactions published articles from a wide array of authors, members and non-members of the Royal Society, scientists and non-scientists. The Journal des Sçavans and the Mémoires de l'Académie des Sciences published almost exclusively work by the members of the Académie (Gross et al., 2002, p. 66-67.).

[35]Morgan 1928; Westfall 1995. De Sallo (1626-1669) was educated in philosophy, classical studies and law. He edited thirteen issues of the Journal des Sçavans, but then seems to have made enough enemies amongst both his authors and the authorities for the journal to be suppressed. It was resumed nine moths later under different editorship. Despite such problems, it was popular enough to appear in translation in Leipzig (1667-1671) and as a counterfeit edition in Amsterdam (1665-1792).

Newton to communicate within a European 'information space'.[36] The last decades of the 17th century saw a limited number of similar ventures published both by the academic societies (imitating the Philosophical Transactions) and private enterprise (following the Journal des Sçavans). Examples include the Giornale de' Letterati (Rome, 1668-1681) and the Acta Eruditorum (Leipzig, 1682-1731). Even so, developments can hardly be called revolutionary, with a mere 20 new and often short-lived titles appearing until the turn of the century. It took until the second half of the 18th century for the scientific journal really to take off, with over 422 titles appearing between 1750 and 1790.[37]

Even a century after the first publication of the Philosophical Transactions the learned journal had not been developed to a stage where it fully met the requirements of scientists throughout Europe. Especially in the case of the academy journals – the journals that could make most claim to being 'peer-reviewed' in terms of the modern concept – there were problems with the time-lag (often a year or more) between the presentation and publication of papers. There was also the problem of language, and the sometimes disturbingly wide array of subject matter. These problems were resolved – coinciding with the decline of the generalist amateur-scientist – through the development of specialized journals that focused on a specific subject matter, presented information from different sources in a single language, and were often capable of much speedier publication.[38] The first of these were in the areas of physics, chemistry and botany, containing original papers, correspondence, translated excerpts of articles published elsewhere, book reviews, etc.[39]

Reflecting the growth of of the new science and its success in establishing itself within the universities, the 19th century saw a significant increase in scientific publishing in terms of the number of journal titles and of published articles. All in all, some 2.000.000 scientific and technical papers were published during the century. This also led to the establishment of a well-organized system for the dissemination of scientific information involving scientific institutions, commercial academic publishers and university libraries. It also created a need for specialized services such as bibliographies, annual reviews, review periodicals, and abstracting and indexing periodicals (a form of

[36]The Philosophical Transactions became popular enough to be translated, appearing in Paris as 'Transactions philosophiques de la Société royale de Londres' (1731-1744). This tradition was continued well into the 19th century, e.g. the 'Årsberëttelser om framsteg i fysik och kemi' published annually by the Royal Swedish Academy of Science from 1822-1850 was also published in German and French translations (Odelberg 1978, cited in Fjällbrant 1997).

[37]Kronick 1976.

[38]Vickery 2000, p. 94, Harmon and Gross 2003, session 2. For a recent overview of the history of chemical journals see Cooke 2004.

[39]E.g. *Observations sur la physique, sur l'histoire naturelle, et sur les arts* (Paris 1771), *Chemisches Journal für die Freunde der Naturlehre* (1778), *Botanical magazine* (1787), *Annales de Chimie* (1789).

'periodicalization' of the bibliography).[40] Also, the practice of including explicit references to previous publications became common around the middle of the century.[41]

The 20th century, finally, saw the well known exponential growth of scientific literature,[42] an increasingly fine-grained specialization leading to the 'twigging' of the scientific journal into specialized titles, the consolidation of the article as the main genre for formal scientific communication, the triumph of English as a scientific *lingua franca* and the establishment – especially through mergers and acquisitions in the second half of the century – of a limited number of large-scale, commercial and highly international academic publishing enterprises, resulting in the increasing importance of shareholder value vis-a-vis the interests of the scientific enterprise.[43] Other characteristics of scientific communication at the beginning of the 21st century include an increase in multiple-author publishing, a decline in the percentage of journal articles read by scientists, declining personal subscriptions, and a continuing importance of older literature as cited in journal articles.[44] Especially in the second half of the 20th century, a significant shift occurred in the relationships between various actors in the information chain:

> 'The relationship between authors, publishers and consumers of scholarly journals was remarkably stable through the 17th to the 19th centuries. It was during the 20th century, after World War II, that the system began to lose equilibrium. Particularly in the past 20 years, the balance struck between the players in the scholarly publication world has been lost'.[45]

Even though there is a noticeable shift towards the United States, Europe remains the home of the leading academic publishers, still capitalizing on the advantages of the initiating role of Europe in the birth of modern science and in the development of world trade and politics in the 19th century.[46]

[40]Vickery 2000, p. 123-124. Abstracting scholarly literature was, however, already practiced in the Middle Ages and was also a feature of early scientific publishing (e.g. in the Philosophical Transactions). One of the first *secondary* publications was the *Jahresberichte über die Fortschritte der physischen Wissenschaften* first published by Berzelius in 1829 (Schofield 1999).

[41]Kim 2001, p. 35.

[42]Increasing by a factor 10 every 50 years over the period 1700-1950 (de Solla Price 1975, Chapter 8) or, perhaps more accurately, doubling about every 15 years (Meadows 1998, p. 15-16). According to Meadows, the rate of increase in the amount of information in scientific journals is now declining, pointing to a flattening off of the logistic growth of scientific information (p. 31).

[43]Mackenzie Owen 2002. Nentwich (2003, section 9.1.3) points to the simultaneous decline of the role of academic and scientific institutions as publishers, and the increasing commodification (i.e. the attribution of commercial value) of scientific information during the course of the 20th century. Kronick argues that although only about 25% of journals were published by scholarly societies during the period studied by him, this amounts to about 50% of all journals published at any given time due to the fact that most other journals were of very short duration (Kronick 1976, p. 121). The commodification of scientific information is possibly related to the commodification of the concept of 'information' or knowledge in general that goes back to pioneers of documentation such as Paul Otlet and Suzanne Briet; see Day 2001.

[44]Liu 2003.

[45]Macdonell 1999.

[46]Newman 1990.

A significant development during the 20th century was the further consolidation of scientific communication as a system for performance evaluation and reputational control as exemplified in the creation of citation indexes, the calculation of impact factors for the evaluation of publications and the formal ranking of journals. Parallel to this the system of peer review, although established in principal from the very beginning of scientific publishing in the 17th century, became codified in its current form of 'editor + two anonymous referees' as a formal distinction between scientific and non-scientific literature.[47]

In the final decades of the 20th century the scientific journal entered a new phase in its evolution in the context of a networked-based 'information society' where information and communication technologies inescapably began to exert their influence on scientific communication – as they do on other areas of human activity. Together with the loss of equilibrium of the communication system, the digitization of the scientific journal has resulted in an increased attention of the academic world for knowledge dissemination as one of its core functions. Much of this attention is focused on the peer reviewed journal as it migrates to the digital environment.

1.5 The development of the electronic journal

As described in the introductory chapter, the beginnings of the 'electronic' journal (or 'e-journal') are to be found in the late 1980's. The development of the electronic journal over the past fifteen years can be described in three overlapping stages. In the initial stage e-journals were characterized by the absence of a printed version. In general these were new, fairly small-scale operations, initiated by an individual researcher or a small group of editors. Distribution was initially by e-mail, ftp or gopher. Most migrated to the world wide web as soon at it succeeded gopher as a structured format. These are the new, 'e-only' journals that at first were expected to set the example for the coming 'revolution' in scholarly communication.

1	1987-	Innovative, e-only journals
2	1997-	Electronic versions of printed journals
3	2000-	Open Access journals

Figure 2.1. The three stages of e-journals

[47]Rowland 2002, Weller 2001, p. 3-8, Meadows 1998, p. 177-194. The importance of peer review for science can be inferred from a large body of research on this topic carried out since 1981, cited by Rowland (p. 249-250).

Around 1997 a second stage set in when existing publishers of scientific journals started to distribute their products in digital form to libraries and over the Internet.[48] In almost all cases the electronic version is identical to the print version,[49] and these dual print/digital products are often referred to as E+P-journals. In a later development these journals became available through intermediary organizations such as Science Direct, EBSCO, Catchword, HighWire, Ingenta, Emerald and others. In a related development backsets of journals became available in digital form, either through the commercial intermediaries or through non-commercial projects such as J-Store.[50] Although the articles in E+P-journals are the same in both versions, there is often a significant difference at the journal or content provider level. Here, the digital version may offer enhanced functionality such as e-mail alerts, user profiles, search functions, cross-linking etc.

A third stage set in around 2000 with the establishment of PubMed Central by the National Library of Medicine.[51] PubMed Central made a first step towards free access to scientific information for the academic community. Under this model publishers deposit their (print and/or digital) journals (at least the peer reviewed content) with PubMed Central, where they can be accessed free of cost over the Internet by end-users. In addition the NLM guarantees the long-term persistence of these journals.

This development is often referred to as *open access publishing*.[52] In a meeting held on April 11, 2003 at the headquarters of the Howard Hughes Medical Institute in Chevy Chase, Maryland, a group of senior individuals from the biomedical research community agreed on the following extended definition:[53]

An Open Access publication[54] is one that meets the following two conditions:

1. The author(s) and copyright holder(s) grant(s) to all users a free, irrevocable, worldwide, perpetual right of access to, and a license to copy, use, distribute, transmit and display the work publicly and to make and distribute derivative works, in any digital medium for any responsible purpose, subject to proper attribution of authorship, as well as the right to make small numbers of printed copies for their personal use.

[48]Peek and Pomerantz, 1998. A well-documented earlier experiment was Elsevier's TULIP project, see Elsevier 1996.

[49]It is interesting to note that initially Elsevier created the digital versions by sending the *printed* journals to low-wage countries, where they were scanned to produce digital images of the original. As a consequence, there was a significant time-lag between the availability of the printed and digitized versions.

[50]http://www.jstor.org/.

[51]http://www.pubmedcentral.nih.gov/.

[52]Björk 2004; Hedlund et al., 2004.

[53]Bethesda 2003.

[54]NB: Open access is a property of individual works, not necessarily of journals or publishers.

2. A complete version of the work and all supplemental materials, including a copy of the permission as stated above, in a suitable standard electronic format is deposited immediately upon initial publication in at least one online repository that is supported by an academic institution, scholarly society, government agency, or other well-established organization that seeks to enable open access, unrestricted distribution, interoperability, and long-term archiving (for the biomedical sciences, PubMed Central is such a repository).

A *commercial* initiative to publish original e-only journals free of charge to end-users was launched in 2001 by BioMed Central.[55] BioMed Central is an independent publishing house currently offering some 100 peer-reviewed biomedical e-journals. As an incentive BioMed Central also provides online submission and peer-review technology without charge to groups of scientists who wish to run open access, online journals under their own editorial control. According to the publisher, 'all the original research articles in journals published by BioMed Central are immediately and permanently available online without charge or any other barriers to access. This commitment is based on the view that open access to research is central to rapid and efficient progress in science and that subscription-based access to research is hindering rather than helping scientific communication.' This development has two important characteristics. One is the focus on new, digital-only journals without a printed counterpart. The other is the use of a new 'business model' based on up-front payment (page charges). Under this model, the author (or the author's institution) pays the publisher a one-off fee to publish an article in a journal. The original research papers published in the journal are then made available free of cost to users worldwide. In addition, the journals may contain value-added content (such as review articles, commentary articles, web reports, image databases and evaluation services), access to which is not free but subject to a subscription charge. As in E+P-publishing, open access journals usually follow the traditional article format, peer-review procedures etc. The difference with traditional journals consists of the new business model and additional functionality offered by the provider.

A further development in the domain of open access publishing was the establishment of the Public Library of Science (PLoS).[56] PLoS started its first open access journal PLoS Biology[57] in October 2003. Its second journal, PLoS Medicine was begun a year later. PLoS signals a shift away from commercial publishing and towards not-for-profit initiatives within the academic community – returning, in a sense, to the practice of phase 1 in the development of e-journals. In addition to its publishing initiatives, PLoS acts as an activist group for open access scientific publishing, based on a statement made in an 'open letter' addressed to the academic community in 2001:

[55] http://www.biomedcentral.com/.
[56] http://www.publiclibraryofscience.org.
[57] http://www.plosbiology.org/.

> 'We support the establishment of an online public library that would provide the full contents of the published record of research and scholarly discourse in medicine and the life sciences in a freely accessible, fully searchable, interlinked form. Establishment of this public library would vastly increase the accessibility and utility of the scientific literature, enhance scientific productivity, and catalyze integration of the disparate communities of knowledge and ideas in biomedical sciences'.[58]

Another example of the development towards open access is the Open Archives Initiative that sets out to enhance access to e-print archives as a means of increasing the availability of scholarly communication.[59] Whether digital archives will replace journals as a means for disseminating research outcomes remains, however, to be seen. The main obstacle to acceptance of digital archives as a mechanism for formal scientific communication is the lack of an embedded peer review procedure. Digital archives or *repositories* do often contain peer-reviewed materials, but these materials have been reviewed in a different context, e.g. through prior formal publication in an established, peer reviewed journal.

Open archives are but one example of a larger array of new genres and dissemination opportunities (including 'self-publishing' options) or 'scholarly communication forums'[60] that have become available to scientists in the digital era.[61] The peer-reviewed scientific journal article, however, whether published in print or digital form, retains its key position as the generally accepted, formal mechanism for the certification, dissemination and archiving of scientific knowledge, the 'canonical form for the communication of original scientific results'.[62] A recent survey of 3.390 UK-based researchers showed that 95% of the respondents regarded refereed journals as essential to their work.[63] The increasing success of open access publishing remains predicated on the 'traditional' genre of the scientific article.[64] Rather than as a transformation of the genre, the move towards open access has to be seen as an attempt to shift control over scientific communication from the commercial publishing industry back to the academic world.

2. The scientific article as a communication medium

As we have seen, one of the important outcomes of the Scientific Revolution was the establishment of a new medium for formal communication: the scientific journal, with the article as the main genre used by scientists to report on their work and document the results of their research. It should be clear,

[58] Public Library of Science (2001), reproduced at http://www.plos.org/support/openletter.shtml.

[59] http://www.openarchives.org/. E-print archives are also referred to as institutional or disciplinary *repositories* (see the discussion of institutional repositories in chapter 8.3 on page 80 ff.)

[60] See section 9 on page 83.

[61] For a recent overview, see Nentwich 2003, chapter 7. See also Mackenzie Owen 2002, 2003.

[62] Gross et al., 2002, p. 4. See also footnote 73.

[63] Education for Change 2002, p. 20.

[64] This issue is discussed in more detail in section 4 on page 167.

however, that our current understanding of what a scientific article is, differs in many respects from what is was a century or three centuries ago. These differences – in language, style, organization, use of illustrations etc. – are immediately apparent if one compares articles from different periods. The scientific article as a *rhetoric genre* has developed over the centuries in a way that requires an analysis that explains how and why scientific thinking and its expression in text and image have evolved towards its present form. Around 1990 the idea that such a type of analysis might inform our understanding of scientific communication led to a number of important, but limited studies.[65] In this section, we shall base our outline of the development of the scientific article on a more recent study by Gross, Harmon and Reidy that is informed by the analysis of a large corpus of English, French and German articles from the 17th to the 20th century.[66]

2.1 The structure of the scientific article

Before outlining the developments of the rhetoric properties, however, we have to describe some general characteristics of the scientific article. First, perusal of the scientific literature shows that the article as an expression of the scientist's work reflects various *different* types of scientific activity (table 2.1).

This typology indicates that it might not always be possible to generalize about *the* scientific article, and that certain properties or developments may be more appropriate for one type than for another.

This applies for instance for a second aspect, the overall structure or 'arrangement' of the scientific article, i.e. the sequence of components indicated by headers such as 'introduction', 'methods', 'results', 'discussion', etc. Over the centuries this structure has become more or less standardized (especially in the more quantitative sciences), but these standardized forms differ across the various types of articles (table 2.2). These different arrangements are, of course, purely functional and the outcome of a selective and consensual process that has developed towards the best contextual solution.

Besides these sequential arrangements, the scientific article has other structural characteristics as well. For instance, the article usually contains a number of formal elements that serve to fulfill specific functions such as the main argumentation of the text (i.e. the content proper), illustration, additional explanation (e.g. through footnotes), external context (e.g. references to the literature) and navigation (e.g. an index). The various functions and their formal elements are summarized in table 2.3.

Finally, the argumentative content of the article itself can be differentiated in various ways. One way is, of course, according to more or less standard

[65] E.g. Bazerman 1988; Prelli 1989; Gross 1990; Pera 1994.
[66] Gross, Harmon, and Reidy 2002.

Table 2.1. A typology of the scientific article

Theoretical	Focus on explanation and conceptual innovation; provide the basis for experimentation and observation.
Experimental	Focus on creation of empirical data through manipulation of objects, usually within the laboratory.
Observational	Focus on creation of empirical data through observation of objects, usually outside the laboratory.
Methodological	Focus on means for performing and improving experiments and observations.
Review	Describes and evaluates published research results in a given field.

Based on Harmon and Gross 2003.

Table 2.2. The 20th century structure of the scientific article

Theoretical articles	**Experimental, methodological or observational articles**
Abstract	Abstract
Introduction	Introduction
Theorem	Materials and methods
Proof of theorem	Results
Conclusion / summary	Discussion
Acknowledgments	Conclusion / summary
References	Acknowledgments
	References

From: Gross et al., 2002, p. 246

Table 2.3. Formal elements of the scientific article

Argumentation	Text body (e.g. structured according to table 2.2)
Illustration	Figures, tables, photo's
Explanation	footnotes/endnotes Glossaries
Context	Bibliographies
Navigation	List of contents List of figures List of tables Headings Marginal text Indexes

classifications of argumentation.[67] A second approach is based on the recognition that the body text of an article is seldom a coherent argumentative structure presenting an idea or research process in a clear-cut way. It often contains many other discourse elements as well that may or may not detract from the central argumentative purpose of the article. Teufel and Moens identify a basic scheme consisting of three elements: background sentences (describing background knowledge), sentences describing the author's work proper (including methodology, outcomes, limitations, further work etc.), and other information.[68] Additional elements include sentences describing the article's research goal, the textual structure of the article, sentences that differentiate the article from other work, and statements referring to work on which the article is based or by which it is supported. Such elements are then combined to form specific discourse structures. Elsewhere we have argued that even formal scientific articles contain, in addition to other elements, a conversational aspect (including even gossip and chit-chat) that should be separated from the content proper (table 2.4).[69]

[67]See for instance Freeman 1991, van Eemeren et al., 1993 and Penrose and Katz (2004) in addition to Gross 1990 and Pera 1994 cited above. See Harmsze 2000 for an application of argumentation theory to the modularization of scientific articles. For an in-depth study of the relationship between academic culture and writing, see Hyland 2000.
[68]Teufel and Moens 1999.
[69]Mackenzie Owen 1989a,b.

Table 2.4. Discourse elements of the scientific article

Informational	fact, theory, law
Procedural	method, instruments
Contextual	description of current state of affairs, problem situation etc.
Argumentational	argumentation, reasoning
Referential	references to other research
Conversational	all other aspects

2.2 The evolution of the scientific article

The first scientific articles were short, reporting on the author's personal experience in carrying out observation and experiment. Baconian scientific practice led to a proliferation of factual data linked together through an informal narrative and with little attempt at explanation. The style of language was everyday prose rather than a specialized 'scientific' language, and was aimed at both amateurs and professionals.[70] Its stylistic and presentational origins in the scientific letter writing of the 16th and 17th centuries are clear.

How are we to describe the evolution from these origins to the sophisticated genre of the scientific journal as we know it today? Gross et al. have analyzed the development of the various structural and rhetoric properties of the scientific article in their study mentioned above. Their analysis focuses on three rhetoric constructs:[71]

- **Style**, including syntactical structure (e.g. in terms of personal pronouns, evaluative expressions, poetic metaphors, passive voice etc.), complexity (e.g. in terms of noun phrases, quantifying expressions, abbreviations and citations) and efficiency (e.g. in terms of average sentence length and clausal density).

- **Presentation**, e.g. in terms of formal structure (cf. table 2.2), graphics and tables, equations, headings, bibliographic information etc.

- **Argument**, e.g. narrative/descriptive *versus* expository/explanatory, qualitative/mechanical *versus* quantitative/mathematical, use of witnesses, observations and experiments, use of graphics *versus* words, etc.

Gross et al. conclude that the scientific article has evolved from the 17^{th}century to the present in the following directions:[72]

[70]Harmon and Gross 2003, session 6.
[71]Gross 1990, p 8-9.
[72]Gross et al., 2002, p. 229-231 and Appendix B.

- The style employed in scientific articles has become more impersonal and objective (following the representation of science as an objective enterprise) at the expense of emotional and literary modes of expression.[73]

- Style and presentation have become more efficient, partly in response to the increasing complexity of science. The greater efficiency is achieved by means of a simpler syntax, a tighter organization and structuring of the content, the use of illustrations, tables, navigational devices (including headings and captions), etc.

- The scientific article has become more uniform worldwide and across disciplines in terms of structure, style and language (English). In terms of structure a limited set of presentation forms and 'arrangements' has emerged in the course of the 20th century, especially in the area of the sciences. These forms enhance the navigational qualities of the article by making it easy to find specific aspects of the research on which the article reports (table 2.2). The scientific article has also become a specialized form in terms of its formal properties, context (e.g. the expected background knowledge of the reader) and target audience (scientific peers, i.e. fellow researchers within the same domain of research).

- Argumentation has become more sophisticated, accurate and precise, employing exposition and explanation rather than narrative; description, quantitative, mathematical and discipline-specific arguments rather than qualitative and mechanical explanations; and a combination of words, mathematical symbols, tables and illustrations rather than words alone.[74]

Gross et et al. summarize the outcome of these developments in a characterization of the current research article as embodying 'a style that represents science as an objective enterprise, fosters more efficient communication, and produces stronger, more flexible argumentative strategies'.[75] In general this style has evolved as a gradual and continuous change over time. However, Gross et al. recognize that in some cases (e.g. in the adoption of the heading abstract in the second half of the 20th century), change can be relatively abrupt. Nevertheless, the scientific article has emerged as a standard form that does not readily

[73]As a result, scientific communication based on the peer-reviewed article serves to transform the subjective perceptions of the author into certified, objective information, and is in fact the indispensable mechanism behind the transition from 'subjective' to 'objective' scientific knowledge as described by Popper (1972, 1978).

[74]This was, however, already a characteristic of the early stages of the Scientific Revolution. The language of the new science was often more mathematical than verbal, and the use of presentational devices such as tables probably also arose out of the more practical needs of applied mathematicians and their clients; see Eisenstein 1980, p. 531.

[75]Gross et al., p. 231.

allow alternative arrangements and deviations from the norm. It is the standard that young scientists early in their careers learn as 'the way it should be done'. Gross et al. ascribe the increasing stylistic and presentational uniformity of the scientific article in part to the standardization imposed by editorial policies and style manuals. They conclude:

> 'So in the latter half of the 20th century, authors of scientific manuscripts face both the general social pressure to conform, present since the very first scientific articles, and the specific pressure to conform to a rule or guideline. Both are in the interest of scientific efficiency; both are at the expense of personal expression'.[76]

It is inevitable that these pressures are a force that should be taken into account when considering the effects of technological innovations on the communications of science and its forms, genres and practices.

[76]p. 174.

Chapter 3

THE SCIENTIFIC COMMUNICATION SYSTEM

1. Scientific communication

The Scientific Revolution created a need for new forms of knowledge exchange that, as we have seen, eventually led to the establishment of the scientific societies and to the birth of the scientific journal. This has to be seen within the context of an emerging research practice that is based on two fundamental concepts: *novelty* and *certification*. Modern scientific research sets out to create new knowledge, adding to the stock of knowledge already available and, if necessary, replacing old knowledge that has been found to be inadequate. In addition, the researcher's own claims as to novelty and validity are regarded as principally insufficient. Therefore research results need to be certified, i.e. subjected to standardized and widely accepted procedures that (at least attempt to) establish the novelty and validity of the researcher's claims in an objective way. Behind these two concepts lies a third concept, viz the idea that science creates – in an abstract sense – a coherent body of knowledge about the world, that is based not on 'opinion' but on 'justified beliefs', that is publicly accessible to the entire scholarly community, and that cannot be manipulated once its contents have been certified. The body of scientific knowledge consists precisely of the cumulative, certified outcomes of research as documented in the scientific literature.

For scientific knowledge to be publicly accessible, it has to be recorded and stored outside of the private realm of the individual researcher. Recording and disseminating research outcomes is also necessary if we accept that certification can only be trustworthy if it is performed outside of the original research environment. Finally, the recorded, documented body of certified knowledge also provides the basis for judging the novelty of a researcher's claims. Therefore, modern science implies that scientific knowledge is not held privately, but is made public, stored and disseminated through a formal, well-organized system that serves to validate and communicate knowledge within the scientific community. It is this system that is generally referred to as the *scientific* (or *scholarly*) *communication system.*

The scientific communication system has a number of specific properties:

- Formal, mediated communication is based on the use of a limited number of recognizable forms (e.g. books and journals) as well as a limited number of specific genres (e.g. monographs (in-depth coverage of a specific subject), 'edited volumes', the research article, etc).
- There is a considerable time lag between obtaining research results and the availability of these results to the scientific community. The time lag depends to a large extent on time needed for (a) writing up research results and submitting the manuscript to a publisher, (b) the certification process, (c) creation of the physical artefact, (d) physical distribution and (e) processing by intermediaries such as libraries. With print publications the cumulative time lag is often 6-12 months for the publishing process alone.
- Within the communication system itself there is no *direct* feed-back possible between author and reader. Feedback is mediated through the same channel (e.g. a 'letter to the editor') and/or form (e.g. an article responding to and citing a previous article).[1]
- Quality control of research results is based on certification of output documents rather than than on certification of the research process itself; in other words, it is based on evidence rather than on protocols.

Quality control is almost universally accepted as an essential and distinguishing characteristic of the scientific communication system. A.S. Duff describes quality control as 'a non-negotionable component of any conceivable de jure model of learned information'.[2] This is even more the case in the context of digital networked information where the abundance of information requires the scientific community to differentiate reliable from uncertified information.

The standard procedure for certification of scientific publications is *peer review.* Peer review is carried out by members of the scientific community within the research field of the author(s). It is an *anonymous* procedure in that the names of the reviewers are not disclosed to the author(s) and (sometimes) vice versa.[3] A further characteristic is that peer review is performed *before*

[1]Outside the formal communication system other, personal forms of feedback are of course possible, e.g. by telephone or e-mail.

[2]Duff 1997, p. 184.

[3]The basic text on quality control in science is Zuckerman and Merton 1971, also published as chapter 21 in Merton 1973. For a concise overview see Parliamentary Office of Science and Technology 2002.

publication and therefore acts as a filter: what is not approved through the peer review process cannot enter the body of formal scientific knowledge.

The scientific communication system has wider social implications within the academic community where it acts as a mechanism for *reputational control*. This works in two ways. One is due to the fact that publication is a means of publicizing the *author:* the dissemination of research results amongst members of the scientific community also spreads the name of the author(s) within that community. The other way that reputational control works is through an elaborate system of ranking and of distributing status within the community. Mechanisms such as peer review, impact factors and citation analysis all serve to create a canonical ranking of scientists according to some measure of 'importance' related to their publications as a representation of their scientific achievements. The scientific communication system is therefore not *only* a system for communicating research results, but also a system that serves to create social meaning and therefore contributes to the social context of science.

In this chapter we begin with a structural analysis of scientific communication, based on a number of models that have been developed to understand and improve the communication process. In a further section we shall discuss the role of the author as an actor in the scientific communication system. We then use our model-based approach to analyze the transition from print to digital formats, and its consequences for the functioning of the communication system. Finally, we expand our formal analysis to put scientific communication in a wider social context, in line with our discussion of innovation and its diffusion in the previous chapter.

2. Models and metaphors: representing scientific communication

How is the scientific communication system organized? In this sectionwe shall describe a number of models that outline its basic components. These models will help to understand not only the way scientific communication operates, but also how its structure has developed over time and how it is influenced by the process of digitization.

Models are abstract representations of reality that we use as a convenient way to discuss and theorize about phenomena. Various types of models can be used for this purpose. Physical reality often can be represented by means of a mechanical model (e.g. of a molecule) or its symbolic representation, e.g. a visual 3D-model created by means of software. Some phenomena can conveniently be modeled by means of mathematical formulæ. But many other phenomena, notably socio-cultural processes, cannot be modeled in this way. Here, complex phenomena are usually modeled by reference to more simple and familiar concepts from a different domain. This type of model was first described by Lakoff and Johnson[4] as a *metaphoric* model: a representation of a (target) domain in terms of another, more familiar (source) domain.[5] The models described in this chapter are all of the metaphoric type.

2.1 The conduit metaphor

Most models of the scientific communication system (as well as most models of communication in general) are based on the so-called *conduit metaphor*.[6] The conceptual metaphor used in the conduit model is that of communication as a 'flow' of information through a 'conduit' or pipeline. This is a model first described by Michael Reddy[7] that is based on the well known Shannon & Weaver transmission model of communication.[8] In conduit models there always is an originating point or 'sender', a 'channel' though which information (i.e. an encoded message) is transmitted, and a destination or 'receiver' where the information is decoded and the message is processed. Fundamentally, this implies that meaning – packaged as information – flows from one point to another, or in our case: from the author to the reader.

The Shannon and Weaver model mainly addresses what Weaver describes as the *technical* problem: the accuracy with which 'symbols of communication' can be transmitted.[9] Weaver describes two further problems of communication: the *semantic* problem (the extent to which meaning is conveyed in a precise way) and the *effectiveness* problem (how to use communication to influence the behavior of the receiver).[10] He indicates that the transmission model and its underlying mathematical theory apply especially to the technical problem of communication, and might not be at all relevant for the other two problem domains.[11]

In digital systems the technical problem is paramount: inaccuracy, even at the level of a single bit, can bring communication to a standstill. In natural

[4] Lakoff and Johnson 1980.

[5] Travers 1996, section 2.1: theories of metaphor.

[6] A very different type of model that we shall not discuss here, is the so-called *cognitive* model. This model views scientific communication as a process whereby information, as a mapping of the mental state of the author, adds new knowledge to the mental state of the reader (Belkin 1990; Ingwersen 1996). This is expressed in Brookes' 'fundamental equation' as: $K[S]+\Delta I = K[S+\Delta S]$ (Brookes 1980). According to this model, adding information ΔI to a knowledge state $K[S]$ results in a new knowledge state that differs by ΔS from the original state. The cognitive model is yet another metaphoric approach, drawing heavily on concepts such as 'adding to the stock' of knowledge. It is likely that Brookes also had an underlying conduit model in mind for the transmission of ΔI to $K[S]$.

[7] Reddy 1993.

[8] Shannon and Weaver 1949.

[9] Weaver 1949.

[10] These three problem areas are often designated as syntax, semantics and pragmatics, or structure, meaning and effect.

[11] In spite of Weaver's cautionary remarks, various attempts have been made to apply mathematical information theory to entirely different domains. For an intriguing example in the area of æsthetic perception see Moles 1958.

language communication, however, accuracy is less important. Although inaccuracy *can* be a serious problem, the redundancy of language usually allows a message to be decoded even if it is severely distorted. Here the problem is much more at the level of semantics. It is at this level that Reddy's conduit metaphor offers a more qualitative view of the human communication process.

Reddy describes human communication as a process in which an actor (e.g. a speaker or author) packages meanings into words that are conveyed by whatever means to another actor (e.g. a listener or reader). There the words are unpacked and the original meaning is recreated. In this model, the problem of communication is no longer the technical problem of transmission but the semantic problem of interpretation. The words received are the exact words used by the sender. The semantic problem is about how the receiver is to derive the original meaning from these words. It is, of course, well known that contextual differences at many levels can lead to mis-interpretation and therefore to failure of the intended communication at the semantic level.

A more structural approach to the semantic level in the context of verbal communication has been developed by Roman Jakobsen. Jakobsen's model describes not only a sender, message and receiver but also a communications *channel* (including the psychological or 'phatic' link between sender and receiver), the *nature* or 'code' of the communication (i.e. the type or genre) and the (shared) *context*.[12] To give an example, this approach would argue that when two scientists communicate, it is relevant that they do so (a) in the context of solving a scientific problem, (b) by means of journal articles and (c) that they recognize and trust each other as competent researchers. Changing any of these three factors would fundamentally change the nature of the communication.

Later models of linguistic mass communication stress its constructivist properties. Notably Stuart Hall has pointed to the distinction between 'encoding' by the initiator of communication, and 'decoding' by the recipient.[13] It is in the act of decoding that meaning is (re)constructed. This distinction implies a communicative space between encoding and decoding in which form and content are available and can be selected and used by the recipient to construct meaning based on a variety of strategies. Similar ideas have been developed by Michael Polanyi, who argues that a message can convey meanings (not necessarily those intended by the author), but not the original experience.[14] This position is summarized by Giovanni Ferrero as follows:

> 'Many of us, when asked about communication, usually think of it as a mere information transmission from a source to a receiver by means of coding-encoding processes.

[12]Jakobsen and Halle 1956; Jakobsen 1960.
[13]Hall 1980.
[14]Polanyi 1969.

However, different studies during the last fifty years clearly underlined that one thing is what is said, another is what is meant. In fact, many meanings are characterized by fuzzy boundaries that are defined by the context in which they are expressed. Within this perspective, meaning is not a fixed system of univocal correspondences between expression and content, but a set of possible inferences that have different probability degrees of realization'. [15]

Most models of scientific communication that adopt the conduit metaphor are based on two assumptions:

1. that there is no specific technical transmission problem: scientific information is assumed to be available in its original, undistorted, authentic form;
2. that there is no specific semantic problem: information is assumed to be comprehensible by the intended audience (i.e. peers within the domain of research).

These two assumptions create the impression of a 'perfect' information environment in science, and tend to downplay potential technical and semantic problems and deficiencies in scientific communication. Even if we accept that these two assumptions hold to a large extent for the traditional communication system based on print publications, it remains to be seen whether this is also the case for digital publications.

2.2 The information chain

The conduit model has often been used as a convenient way to describe the practice of scientific communication. Ideas (whether or not based on the outcomes of research) are 'packaged' into specific forms (e.g. the journal article). These packages (usually bundled into journals) then flow through an organized system (incorporating quality control, storage and retrieval mechanisms) to the user at the far end of the conduit, adequately described as the 'end user'.[16] This conduit is generally seen as a complex system, a sequence of steps through which information passes, each step being carried out by different actors performing functions specific to that step. The conduit for scientific information is often referred to as the scientific (or scholarly) *information chain*.[17] However, this is not in all respects an accurate representation in view of a number of characteristics that we shall describe in this section.

The basic structure of the information chain model consists of the author and the reader at the outer ends of the chain, with a limited number of institutional intermediary actors (e.g. publishers and libraries) in between (fig. 3.1).

[15]Ferrero 2001.
[16]The end, that is, from the perspective of the author, but also because of the time sequence inherent in the conduit model. The author stands at the beginning of the sequence, the user at the end.
[17]Duff 1997.

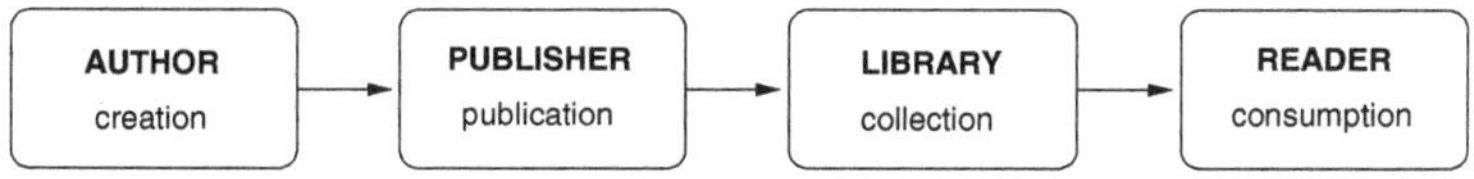

Figure 3.1. The information chain

This simple representation links specific functions (e.g. creation, publication, collection and consumption) to the various actors.

A variant of the conduit model as applied to the scientific communication chain is the concept of a *value chain* that models communication as an economic activity consisting of a chain of linked, interdependent activities. The idea is that each component in the chain adds value to the message or 'information product' in some way.[18] Roosendaal c.s. use this concept to describe structural changes in scientific communication induced by the use of information and communication technology.[19]

This type of model raises two key issues:

1. How is the information chain organized? What are the various actors and functions, how are they interrelated, how do they develop over time?

2. Given that the volume of communication exceeds the capacity of the individual end user, how can he/she select pertinent messages without reviewing them all (the *information retrieval* problem)?

The first question has been answered in different elaborations of the information chain model, as we shall see later on in this chapter. The second key question – that we shall not elaborate on here – illustrates the fact that the scientific information chain is in fact *not* a pure conduit system. Firstly, there is no direct flow of information from the author to the end user. Instead, the information chain contains a *memory* or repository in which messages are deposited and from which they are drawn if and when they are requested by an end user. In analogy with many systems in physical logistics which include warehouses, reservoirs etc., this type of communicative system can be described by means of a *repository metaphor*.[20] This metaphor allows further analysis of the scientific communication system in terms of a clearinghouse[21] and market place (see fig. 3.2 where the library performs a repository function as a clearinghouse).

Secondly, and more importantly, the information chain is also a system in which the user actively engages himself by utilizing functionality embedded in

[18]Franstvag 2002, p. 3, Hedlund et al., 2004, p. 201.
[19]Roosendaal and Geurts 1997; Roosendaal et al., 2001.
[20]See also p. 78 ff.
[21]A clearinghouse (Marron 1971) is a method for reducing the number of links or interrelationships between two sets of social actors, e.g. between publishers and readers.

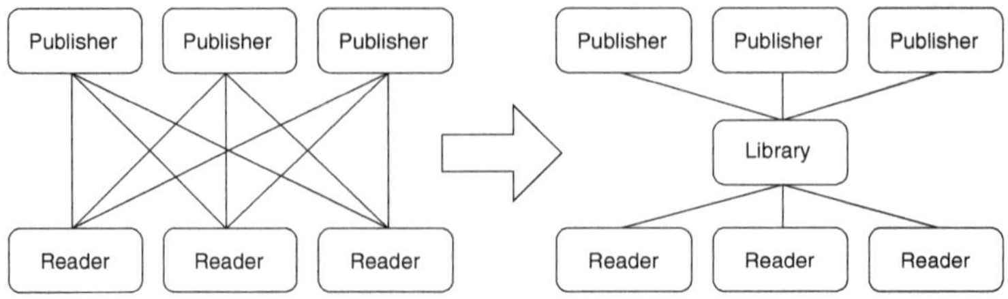

Figure 3.2. The library as a clearinghouse

the system, by selecting relevant information entities, and by constructing, at a macro level, an aggregated view of the research domain by means of a personal selection of information resources. We propose the term *engagement model* for this interpretation of the information chain as a more active system comprising a certain level of user involvement. The engagement model is related to Reddy's concept of the 'toolmakers paradigm'.[22] In this paradigm, based on a constructivist theory of language, messages are perceived as 'instructions for the construction of meanings' rather than the carriers of meaning itself. The toolmakers paradigm implies a higher level of engagement from the user than within the conduit metaphor, and accepts as the outcome of the communicative act different results for different end users.

On further inspection it appears that the 'information chain' is a dual concept that can be described by a conduit model *and* by an engagement model, but at different levels:

- The *conduit model* applies to communication at the micro-level of the individual article (or other communicative genre): here a message is packaged by the author, transmitted in a neutral way to the end user,[23] and unpackaged to acquire the original message.

- The *engagement model* applies to the macro level of scientific communication within a specific domain: the end user is not a passive receiver of individual, isolated messages, but one of many actors who construct their individual views of the scientific state of affairs – and also collectively as the 'prevailing opinion' on a certain issue – by selecting and aggregating a larger number of appropriate messages.

3. Early models of the information chain

Most representations of the information chain are based exclusively on the conduit metaphor in that they describe the flow of scientific information from

[22] Illustrated in Reddy 1993, fig. 1.
[23] Neutral, that is, if we accept the published document as the author's 'statement'.

authors and research organizations to other research organizations and readers.[24] While authors and readers are treated more or less as 'black boxes', much attention is given to detailing the organization of the intermediary processes in terms of functions (what is done) and actors (by whom).[25] Some models also include the genres or informational forms that are utilized within the information chain, i.e. the various forms of primary (manuscripts; books and journals), secondary (e.g. bibliographies) and tertiary (e.g. bibliographies of bibliographies) literature.[26] A characteristic of some later models is that they acknowledge the *cyclical* nature of the information chain due to the fact that authors and readers belong to the same community and are, at the conceptual model's level of abstraction, the same actor.

Although some earlier models have been published,[27] the first concerted effort to create a formal model of scientific communication was the *UNISIST* model created in cooperation by the United Nations Educational, Scientific and Cultural Organization (UNESCO) and the International Council of Scientific Unions (ICSU).[28] This model is loosely based on a typology of genres and the various actors involved in processing these genres. For the formal, published literature the model describes the classic information chain (producers. publishers, libraries and users as producers and books and journals as primary genres), supported by various secondary genres and services for abstracting, indexing and data handling (table 3.1). The model is important in that it introduced concepts of computerized data processing at an early stage.

Perhaps the best known of all attempts to provide a conceptual model of the scientific information chain is that by F.W. Lancaster.[29] His cyclical model identifies the key actors and their roles, expressed as functions in the communicative process (fig. 3.3 and table 3.2).[30]

Lancaster's model is interesting in that, although it is cyclical, it is not so in the sense outlined above. Lancaster still distinguishes between authors and readers with their respective roles ('writing' and 'assimilation'). But the pivotal point in which beginning and end of the communicative process come

[24]For an overview of of some more recent information chain models see Duff (1997).

[25]Relatively little is known about how researchers package their ideas into informational forms such as research articles, nor about how they unpackage these forms to re-create meaning. In communication studies and especially within information science the main issue is the conduit itself, the link *between* the researcher-as-author and the researcher-as-reader. Information science has also provided a large body of research on the information seeking behaviour of readers (notably scientists), but very little on how information is processed once it has been acquired. One of the few authors to have developed a specific view on the role of information within the research process is Latour (1987).

[26]See for instance Subramanyam 1979.

[27]Urquhart 1948; Judge 1967.

[28]UNESCO/ICSU 1971. See also Søndergaard et al., 2003 for a revised update of the UNISIST model.

[29]Lancaster 1978.

[30]It should be noted that what Lancaster designates as an 'information centre' is, in a broad sense, what most people would describe as a library.

Table 3.1. Information genres in the UNISIST model

	Published	**Unpublished**
Informal	Letters to editors, preprints	Talks, lectures, conferences Theses Reports
Formal	Books, journals	
Tabular	Quantified surveys	
Secondary	Abstract and index journals Special bibliographies Reviews, syntheses	Catalogs, guides

Table 3.2. Functions in Lancaster's model

Author	R&D activities writing
Publisher	publication distribution
Information centre	acquisition organization and control archiving arranging cataloguing classifying indexing
Users	assimilation

together, is described by Lancaster as *research and development activities,* i.e. as a separate process that lies outside the domain of the communication system. This suggests that Lancaster does not regard the communicative acts of writing and reading (or 'assimilating') as belonging to the domain of research itself. Also, Lancaster's model does not differentiate between various communicative genres or disciplinary domains. This is unfortunate, as for this reason Lancaster's influential model has tended to impose a monolithic conception of

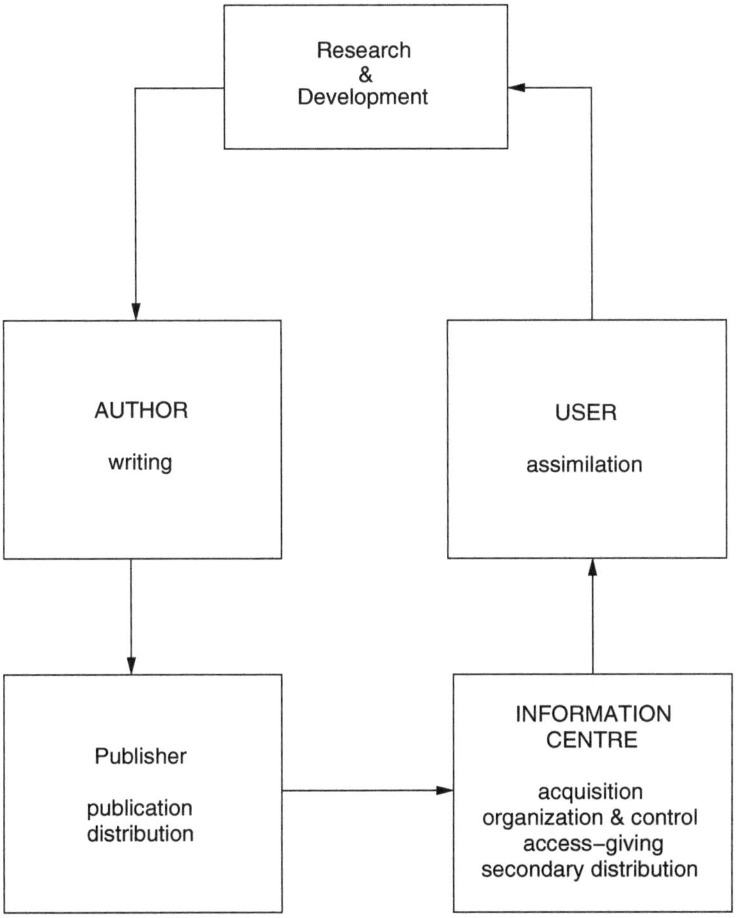

Figure 3.3. Lancaster's model of the information chain

the information chain on the discourse about scientific communication.[31] Another objection to the model is that, although it depicts sequential stages in the communication process, it offers no indication of the time scales.

Another influential model, however, did introduce some notion of genres as well as a clear consideration of time scales. This is the model developed by Garvey and Griffith (fig. 3.4).[32] Although allowing for informal and peripheral genres (e.g. preliminary reports, seminar presentations, conference

[31]That would appear to explain why Søndergaard et al., returned to the UNISIST model rather than to Lancaster's model as a background to analysing and comparing different knowledge domains and their communication structures.

[32]Garvey and Griffith 1972; Garvey 1979.

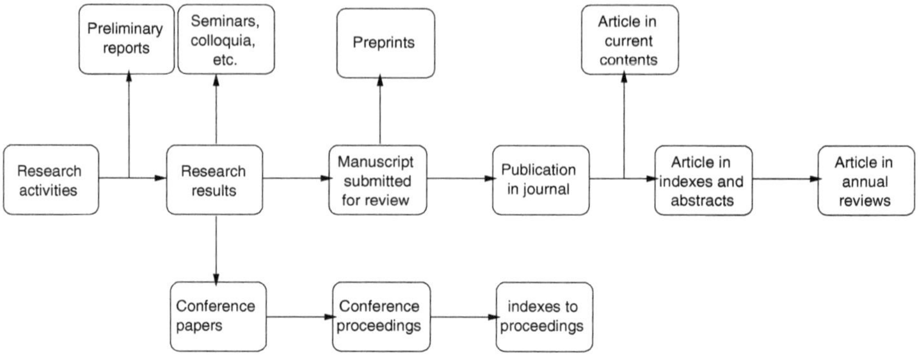

Figure 3.4. Garvey-Griffith model of scientific communication

papers, preprints, indexes and reviews), the core genre of the scientific communication process is firmly established as the scientific article published in a peer-reviewed journal. The model acknowledges different information chains for different genres and identifies various inter-genre relationships.[33]

Based on their analysis of extensive datasets, Garvey and Griffith were able to establish a general timeline that shows how scientific work first finds expression in the initial scientific article and then slowly permeates into other articles and genres (table 3.3).

4. The information chain as a transaction space

Most conduit models of the Lancaster type are based on a deterministic view of scientific communication. They imply that whatever an author introduces into the system will eventually, after a fixed number of steps, reach the end user. In practice this is not at all true: the information chain is not a conduit where what goes in must always come out. Some publications will not be noticed, some will be neglected, and it certainly is not the case – fortunately – that each and every publication will flow to each and every end user. These models are also specifically unidirectional and distribution-oriented and can be described as 'push-models'. They do not allow for an acquisitions-oriented view (or

[33]Kling, Spector, and McKim (2002) point to differences across disciplines, especially in nomenclature:

> 'While many scholars believe that the trajectory of publication described by Garvey and Griffith fits many fields, there are important variations in sequence and nomenclature across disciplines. These differences in the nomenclature for research articles, such as, preprints by high energy physicists, working papers, memoranda, research manuscripts, and technical reports by others, continue today.'

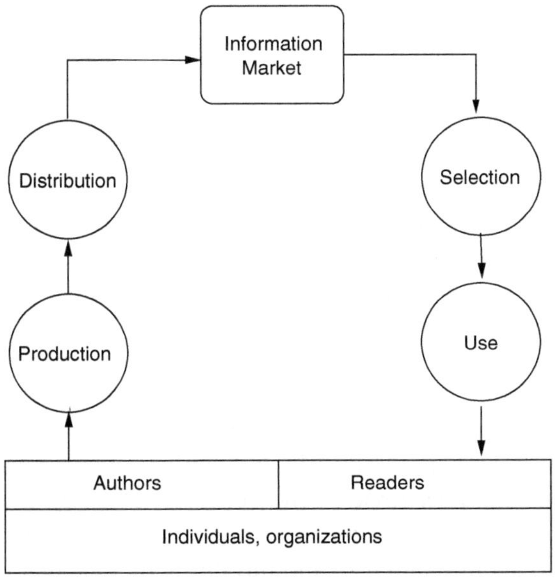

Figure 3.5. Scientific communication as a market for information

Table 3.3. Time scales in scientific communication

After # years	
0	Start of research
1	Completion of research
1.5	Conference paper
2	Submission of manuscript
3	Publication in journal
4	Included in abstract and indexing services
5	Cited in annual review
7	Cited in other articles
8	Cited in other review articles
15	Cited in specialized texts, treatises etc.

'pull-model') in which communication is initiated by the reader. To provide a more realistic representation of what actually happens in the information chain, we take a different approach here (fig. 3.5).[34] This approach takes individuals and organizations as its anchor point, differentiating between their roles of authors and readers.

[34]This approach was originally suggested in Mackenzie Owen and van Halm 1989, p. 1-7.

The model defines two author-related functions (production and distribution) and two reader-related functions (selection and use). Production refers to the production of a marketable information product on the basis of the author's research output. Selection refers to the various activities performed by the user to identify, find, evaluate and acquire available information products. In contrast with other models there is no direct, sequential link between author and reader. Instead, the model defines a central 'information market' where authors (as represented in their information products) and readers interact. The concept of a market place allows for 'extra-scientific' concepts such as marketing, pricing and competition within the context of scientific communication.[35]

An important aspect of this model is the notion of *selection* as a factor in the information chain. The model illustrates the fact that selection is not only an issue at the beginning of the information chain where authors decide what to write and how and where to publish, and publishers select papers on the basis of quality (through peer review) and commercial criteria. At the 'far end' of the information chain, selection is as much an issue, where readers decide what they wish to acquire and use, and through which channels. Traditionally both types of selection have been seen more as functions of intermediary actors rather than as related to the author/reader: publishers decide what is made available in the marketplace, while libraries decide what is to be passed on to users, and how.[36] In fact, however, there are two information markets within the information chain. One is the 'traditional' trade market between publishers and libraries. The other is the more conceptual market where authors offer and readers select and acquire research information and ideas through the medium of the scientific article, packaged in journals and transmitted by the publisher/library complex.

In a broader sense we propose to use this model as a representation of the information chain as a *transaction space,* i.e. as a space where scientists (in their roles as authors and readers, and mediated through publishers and libraries respectively) negotiate issues such as quality, prestige, scholarly and social recognition, as well as more material issues such as price and financial rewards in the exchange of scientific information.[37]

The idea of a transaction space in which it is ultimately the user who controls whether or not a communicative transaction will be performed, underlies

[35] Interestingly, these are not concepts that are universally welcomed in discussions of the information chain. Duff (1997, p. 185) describes the concept of the 'information market' as having '... strongly commercial and capitalistic connotations', and as an 'arresting feature (that) may be a weakness of the model'. The representation of scientific communication as a value chain (Roosendaal and Geurts 1997; Roosendaal et al., 2001) is also predicated on the concept of a market for scientific knowledge.

[36] This is a key point for advocates of self-publishing models, who argue that digitization has the potential to shift de decision power towards the extremes of the information chain, towards authors and readers themselves.

[37] The view of the information chain as a transaction space is related to Stewart Hall's constructivist approach to communication and Reddy's 'toolmakers paradigm' – see p. 2.1 and 3.2.

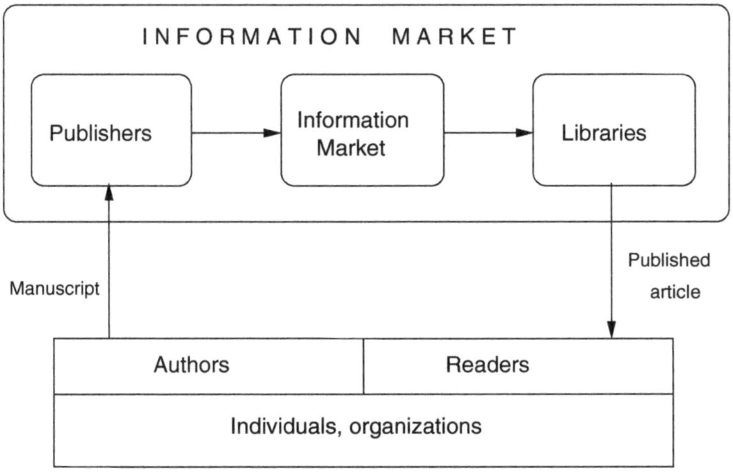

Figure 3.6. The dual market of scientific information

the dimensions developed by Kling and McKim for effective electronic journals: publicity, access, and trustworthiness.[38] If the user does not know about a publication, is not able to access it, or does not trust it, a communicative transaction does not occur. Importantly, Kling and McKim stress the role of peer review for the trustworthiness of research publications. What this amounts to is that quality control is not only essential to science in an abstract sense (i.e. as an appropriate guarantee against mistakes, fraud and deception), but also as a social (and psychological) precondition for communication. Peer review operates therefore both as a selective filter – effectively blocking access to unacceptable information – and as an enabler of trust without which communication would break down.

Hummels and Roosendaal elaborate on the role of trust in scientific communication, distinguishing as we do here between author-related and reader-related functions.[39] They argue, expanding on a model developed by Zucker,[40] that trust is a multi-valued concept consisting of:

- *Process-based* trust: based on shared experience in performing transactions
- *Institutional-based* trust: based on formal structures, procedures, verification etc.

[38]Kling and McKim 1999.
[39]Hummels and Roosendaal 2001. They define registration and certification as author-related functions, archiving and awareness as reader-related. See also section 6 on page 64.
[40]Zucker 1986.

- *Characteristic-based* trust: based on shared characteristics, values and principles and a common background

- *Values-based* trust: based on membership of a community of stakeholders with shared objectives[41]

We can relate these different dimensions of trust to scientific communication as follows. Process-based trust derives from the established scientific publishing procedures with which scientists are commonly familiar. They share experience of 'the way the system works' with their peers. Institutional-based trust derives from various formalizations that are embedded in the communication system. This includes, of course, peer review, but also the formal structure of the scientific article as outlined in the previous chapter. Characteristic-based and values-based trust derives from the socializing role[42] of the body of scientific literature within the discipline. This role serves to define specific values and to establish an 'invisible community' of scientists with a common background and shared objectives.

Hummels and Roosendaal conclude from their analysis that the development of scientific communication has shifted the focus from process-based trust towards institutional and even characteristic-based trust. They also expect future developments, notably digitization, to require a further shift to characteristic and even values-based trust, e.g. in the light of e-science developments such as real-time information exchange during the research process.

5. Towards a continuum model of the information chain

Since the conduit metaphor is a sequential representation containing a time dimension, it can also be viewed as a *life cycle* model that describes the various stages that a document traverses in its passage through time from creation to obsolescence (or at least long-term archival storage). The drawback of this view in a theoretical sense is that it describes the various stages of the communications process in isolation, as discrete steps where functions are performed with little reference to previous or succeeding steps. This life-cycle view does, however, reflect the current reality of the information chain where there is no overall control over the process of scientific communication.[43] Each step is governed by an independent actor (author, publisher, library) with little or no reference to common goals. The result is a loosely coupled and locally

[41]The difference with characteristic-based trust is that in values-based trust participants in the transaction space may have very different characteristics, but nevertheless share a common goal and therefore be willing to engage into a transaction.

[42]Brown and Duguid 2000.

[43]Mackenzie Owen 2002, p. 276.

optimized system that cannot easily adapt to major external pressures or internal conflicts.[44]

A more integrated model describing the interdependencies within the information chain would be useful to describe a future situation where a higher level of coordination might be applied. An example of this type of approach is the *records continuum model* that has been developed in the domain of archiving.[45] The concept of the 'records continuum' is defined by the Australia Records Management Standard AS4390 as 'a consistent and coherent regime of management processes from the time of the creation of records (and before creation, in the design of the recordkeeping system) through to the preservation and use of records as archives'.[46] The continuum idea implies therefore some form of overall control over a process in terms of its components as well as over time. In terms of scientific communication, this would imply the academic community exercising more control over the entire process of communication than is presently the case. It would also imply that the process of creating scientific information should encompass issues of dissemination, retrieval and use as well. The records continuum model can be seen as a critique of traditional life-cycle based approaches and their perceived tendencies to lock actors into specific roles, focus on the object rather than on functions, and neglect the need for a 'holistic' responsibility and management of information processes.[47]

A recent model of scientific communication that goes in a similar direction has been developed in the context of the SciX project funded by the European Commission.[48] This project aims to develop new, more cost and time efficient business models for the scientific publishing process. This objective is based on the assumption that 'the scientific publication process has been so far only marginally affected by the possibilities of the Internet [due to] a lack of sound business models and pilots to demonstrate the ultimate benefits of free scientific publication'. One of the outputs of the project is the Scientific Publication Life-Cycle Model.[49] The model describes the life cycle of a refereed scientific article from its inception to its later use (i.e. being read or its contents being applied).[50] Its main objective is to support the studying of the cost implications

[44]An example of the latter is the so-called *serials crisis,* cf. Cummings et al., 1992; Kaufman 1998; Mobley 1998.

[45]The model – essentially an Australian concept developed at the Department of Librarianship, Archives and Records at Monash University – was originally formulated by Frank Upward (Upward 1996, 1997), based on ideas developed by Ian Maclean and Jay Atherton. The model is sometimes incorrectly attributed to David Bearman, see Upward 2001, note 1. For a further discussion of the development of the model see McKemmish 2001.

[46]AS4390 1996, part 1: clause 4.22.

[47]McKemmish 1997.

[48]http://www.scix.net/. For the final results see Turk 2004.

[49]Björk et al., 2002; Björk and Hedlund 2004. The term 'life-cycle' is somewhat misleading in that it refers to the organizational concept of a continuous and integrated process

[50]Note that the life cycle in this model does not address the concept of a document being destroyed, the assumption being that scientific information will be archived indefinitely.

A0 Do Research, Publish, Study and Exploit the Results

A1 Perform the Research

- **A2** Publish the Results
  - **A21** Write Manuscript
  - **A22** Perform Publishing Activities
    - **A221** Publish as Monograph
    - **A222** Publish as Conference Paper
    - **A223** Publish as Scholarly Journal Article
      - **A2231** Do General Publisher's Activities
      - **A2232** Do Journal Specific Activities
      - **A2233** Do Article and Issue Specific Activities
        - **A22331** Article Specific Activities
        - **A22332** Prepare Issue
        - **A2333** Publish Article
    - **A224** Publish in Miscellaneous Form
  - **A23** Archive and Index
    - **A231** Make Publication Available
      - **A2311** Secure Access Rights and Subscription
      - **A2312** Make Paper Publication Available
      - **A2313** Make Electronic Copy Available
      - **A2314** Integrate Meta Data into Search Services
    - **A232** Perform Value-Adding Services
    - **A233** Archive Securely
- **A3** Study the Results
  - **A31** Find out about Publication
    - **A311** Search for Publication
    - **A312** Be Alerted to Publication
  - **A32** Retrieve Publication
  - **A33** Read Publication
- **A4** Implement the Results

Source: Björk and Hedlund 2004, p. 11.

Figure 3.7. Life cycle model – hierarchical structure

of various business models (e.g. open access / up-front payment *versus* commercial / subscription-based). Interestingly, the model does not address the properties of the scientific journal itself but focuses on a large number of operations that are carried out by various actors as they perform their respective functions. These operations include writing, editing, printing, distributing, archiving, retrieving and reading the article (fig. 3.7). In this respect it describes the *workflow* of the publication process rather than the life-cycle of the scientific publication. This type of modeling remains, however, a theoretical exercise as long as it does not form the basis of coordination and higher-level control within the information chain. Suggestions in this direction have been made by Buck c.s. of the California Institute of Technology by proposing a new administrative model based on enhanced responsibility for and control by the academic community over the process of scientific publishing, with a consortium of universities as its main organizing body.[51]

6. Functions of scientific communication

Lancaster's model of scientific communication has caused considerable discussion – notably amongst librarians and publishers – about the exact nature of the *functionality* of the information chain and how the various functions should be attributed to the actors involved in scientific communication.[52] A concise and influential outline of these functions has been given by Rowland in a 1997 article about the future of the print journal.[53] He identifies three key functions of scientific communication above and beyond the obvious function of information *dissemination.* The first of these is *quality control* as performed by the peer review process, essentially a selective filtering that regulates access to the information chain. The second function is described by Rowland as the *canonical archive.* This concept refers to the fact that the aggregate of scientific publishing constitutes the record of research outcomes at least from the mid seventeenth century onwards. Rowland seems to suggest that without a well-organized information chain research outcomes still could be disseminated, but that there would probably not be a formal and standardized record accessible to the entire scientific community. Finally, Rowland acknowledges the social role of scientific communication in that it provides *recognition* and confers status to the scientific author.

Rowland appears to imply that these functions are related to the way the communication system is organized (especially the role of the publisher), and goes on to discuss the consequences of digitization for academic publishers. However, it is useful to make a distinction here between functions that are directly related to the actor, and functions that are more related to the communicative act and the underlying information genres. Distribution and quality control are functions that must be *organized*, they result from purposeful actions by institutionalized actors. Here the role of the publisher as actor is clear. But the canonical archive and status are not 'organized' in this sense, they 'come about' when scientists make their research results public and these

[51] Buck et al., 1999.

[52] See for instance Cox (1999) and the ensuing e-mail discussion beginning at http://www.library.yale.edu/~llicense/ListArchives/9908/msg00004.html. See also Consortium for Educational Technology for University Systems 1997 for a discussion of library roles.

[53] Rowland 1997

results are accepted as valid and are 'assimilated' (to use Lancaster's expression) by their peers. This can happen whether or not actors such as publishers and libraries are involved. We therefore prefer to call these the *intrinsic* functions of scientific communication. These intrinsic functions are intimately related to the choice of genre, because different genres are interpreted and valued differently by the scientific community.

To these intrinsic functions of the document one should certainly add that of *expression,* i.e. the basic articulation of research outcomes. Another intrinsic function, noted by Schaffner for the scientific journal and by Brown and Duguid as a more general property of documents, is the capacity for *building communities* of participants with a shared context (e.g. of interests, terminology and reputation) – itself, as we already have noted, a pre-condition for effective communication.[54]

A slightly different approach to the functionality of scientific communication is described by Kircz and Roosendaal in the context of electronic publishing.[55] They distinguish between *certification* (i.e. the validation of research quality, related to the scientific standards within a research programme), *registration* (allowing an individual scientist or research group to claim priority for the research outcomes), *awareness* (i.e. publishing not just as 'making public', but also as publicizing), and the *archival* function. Hummels and Roosendaal relate these functions to their analysis of various types of trust.[56]

If we now take together various views on roles and functions in the information chain we arrive at the concise overview given in table 3.4.[57] Here we see how in the system for scientific communication, as it has developed over the centuries, different actors have consolidated specific functions, while on a meta-level the system is characterized by a number of intrinsic functions that cannot be ascribed to individual actors.

We have pointed out elsewhere that the attribution of functions to actors in the information chain is no longer stable.[58] Authors have been forced to take on a number of functions traditionally performed by publishers, especially in the area of editing and typesetting (e.g. the requirement to produce 'camera-ready copy'). There is even some discussion about 'self-publishing' by academic authors and their institutions.[59] Publishers have been taking on library-related functions such as cataloguing and indexing, archiving and end-user services

[54] Schaffner 1994; Brown and Duguid 2000.

[55] Kircz and Roosendaal 1996. This approach is related to Kaufer and Carley (1993) who distinguish ownership of an idea, societal recognition for the author and priority claims for discovery as important factors in academic writing.

[56] Hummels and Roosendaal 2001, p. 97-98. See also this chapter, p. 59.

[57] Note that this list is, as any model, a simplification. For instance, we have omitted for reasons of efficiency the role and functions of subscription agents that operate as clearinghouses between publishers and libraries.

[58] Mackenzie Owen 2002, p. 278-9.

[59] Hibbitts 1999; Jobson 2003; Kling et al., 2002.

Table 3.4. Functions in the information chain

Intrinsic	Expression Reference Canonical archive Recognition / status Community-building
Author-related	Research Writing of publications Editing Reviewing
Publisher-related	Registration Quality control / certification (selection, organization of peer review) Contextualization (series, journal titles, cross-linking) Editing and lay-out Marketing / awareness and dissemination Author support
Library-related	Selection and acquisition Contextualization (collection) Cataloguing and indexing Storage / archiving Retrieval/access and delivery services User support
User-related	Searching, selecting and acquiring Reading Incorporation in research and publications Translation of research results to practical context Application of research results

such as document delivery. Libraries, on the other hand, have been adopting publisher-related functions by creating information products such as electronic journals.[60] The role of digitization is apparent in this development in that it seems to concentrate functionality in a single point around the digital artefact. The result is a struggle for control over the artefact and artefact-related functions amongst the various actors. This struggle lies at the basis of new business

[60]For instance through the FIGARO project (http://www.figaro-europe.org), see Savenije 2003.

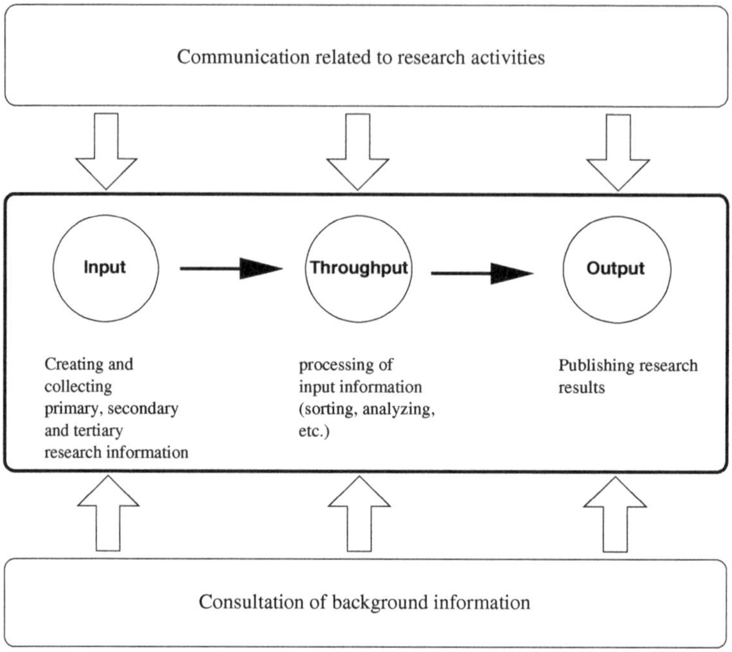

Figure 3.8. A 3-phase model of research

models for scientific publishing (e.g. open access publishing) described in the previous chapter.

7. The author as an actor in the information chain

Our main interest in this study lies in the impact of digitization on form, function and content of the scientific article as a communicative genre. Since it is the author who ultimately defines how digital means are employed in the creation of an article, we now turn our attention to the role of the scientific author as an actor in the information chain. To clarify the various roles of the researcher we use a simple phase model to describe the information-related activities on which these roles are predicated (fig. 3.8). This model identifies three distinct phases in the research process:[61]

- An *input* phase where the researcher gathers and selects the data or information that is to be used for carrying out the analytical research process proper. To a certain extent, and in differing degrees depending on the discipline and type of research, the research data may be *generated* within the

[61] This model is – as are most sequential models – an abstraction. In practice phases may overlap, or a research project may consist of a larger number of overlapping input–throughput–output cycles.

research environment itself, i.e. by means of laboratory experiments, data sensors, interviews etc. We refer to this type of data as *internal* research data since they are created by and 'belong to' the research activity, and may not yet have been made public. A second type of data we refer to as *external* since it consists of publicly available, pre-existing sources such as archival materials, datasets, 'grey literature' and published scientific literature.

- A *throughput* phase where internal and external information is processed and analyzed to produce research results. Multiple iterations of input and throughput are of course common before this phase is finalized.

- An *output* phase where information (in the form of reports, conference papers, journal articles etc.) on the research activity and its results is produced and made public or 'externalized'. This information is then available as input to other research activities.

In addition to identifying an internal/created and an external/consulted information domain, the model also identifies a third, interactive information domain that is concerned with informal *communication.* Scientists generally do not work in isolation, but maintain many diverse contacts with colleagues and resource persons with whom they exchange information through a large array of communicative forms (face-to-face, telephone, e-mail, discussion lists, conferences etc.). This form of information exchange is informal, non-certified and often private, and is an ongoing activity throughout (and beyond) the research process.

It is important to understand that all three phases described by the model are components of the research activity and that the input and output phases should not be regarded as peripheral. Also, although the model suggests clear distinctions and sequentiality, this is not always the case. For instance, there may be numerous alternations between input and throughput, and/or between throughput and output before a research entity is concluded, e.g. by the publication of a research article. We postulate that any research activity begins with the consultation of public information as a preliminary to defining the research problem, organizing data collection, experimental setups etc. In general the consultation of public information will decrease during the throughput phase (as the focus comes to bear on internally generated data), but might increase again at the output stage where external sources are sought in the writing of the research outputs. It is probable, however, that there are different patterns in different disciplines. For instance, it can be expected that the input phase is more distinct in the sciences than in the humanities, where the consultation of external sources is a more continuous and integrated research activity. In the humanities the emphasis typically is on the throughput phase while the input

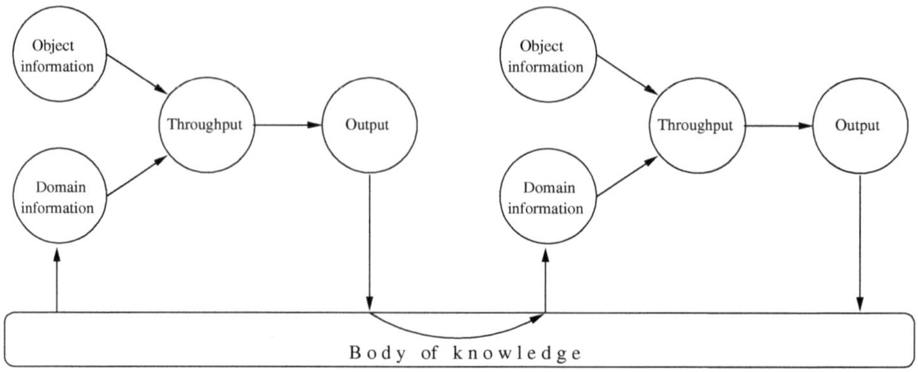

Figure 3.9. The extended 3-phase model

phase may be relatively simple. If input information (an archive, for instance) is readily available, the burden of work consists of making sense of that information (throughput) and creating a research output. In other forms of research (e.g. in the sciences and often also in the social sciences) the burden of work may consist of creating the input data, for instance by carrying out lengthy and complicated experiments. In such cases analyzing the data (throughput) and writing up the research results (output) may be relatively trivial.

The various elements of our model can be characterized in terms of a communications typology that distinguishes between *conversation* (exchange of information between individuals), *allocution* (exchange of information from one to many), *consultation* (accessing available information sources) and *registration* (creating or adding to an information resource). The Input Phase of our model is a form of consultation, based on acquiring information on the existing body of knowledge and/or on on the object of study. The Output Phase is based on a combination of allocution (reporting research results to a larger group of scientists) and registration (adding to the codified body of knowledge). Conversation is, as mentioned above, an ongoing activity that permeates the entire research process.

The research model described in fig. 3.8 can also be represented in a different way that makes explicit the distinction between input information that is related to the research object, and input information that is related to the body of knowledge available within the research domain (fig. 3.9). This version of the model points to the complexity of the external body of knowledge that is consulted by researchers and to which they contribute. It consists not only of the formal, peer reviewed journal literature and informal and unpublished resources, but also many other types of accessible information inputs: archival resources, datasets etc.

A slightly more elaborate model of the research process has been described by Jerome Ravetz, outlining the various types of information relevant to the

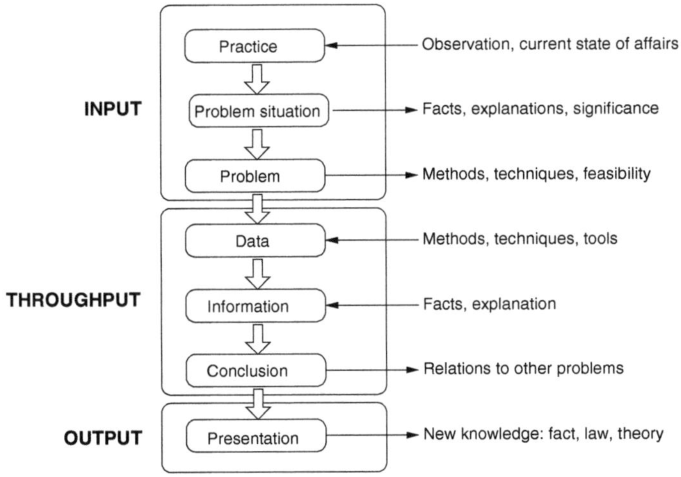

Figure 3.10. Elaboration of the research process (partly based on Vickery 2000, fig. 12)

different stages of research.[62] If we relate these to our 3-phase model (fig. 3.10), we see that the input phase is concerned with surveying the problem area, understanding what is already known (in terms of available data and explanations), and assessing the availability of research instruments (methods, techniques, tools) and the feasibility of the project. The throughput phase is mainly concerned with the transformation of input data to meaningful information and to drawing conclusions with respect to the research problem. The output ideally consists of new knowledge as facts, laws or theories.[63]

Our theory that the research process consists of three distinct phases is supported by data compiled by Garvey on the use of information resources by researchers.[64] Garvey's findings, based on a rather different classification of research stages (see table 3.5 where we have related Garvey's categories to our three phases), show a sharp drop in the use of external information sources during the throughput phase (fig. 3.11). There is also a marked difference between the input phase where there is a balance between formal and informal sources, and the output phase where the emphasis is more on informal, personal sources.

A significant property of the traditional research process is the discrete (i.e. non-continuous) nature of its *outputs*. Usually the researcher(s) will spend a

[62]Ravetz 1971.

[63]Ravetz' model of the research process presents an ideal case, abstracting from what we have defined as communication related to research activities (the informal exchange of information between scientists) as well as from procedural, contextual and conversational components of research articles.

[64]Garvey 1979. Garvey makes the distinction between *personal* sources (i.e. informal communications, either face-to-face or though attendance at conference etc.), *research reports* (formal communications such as journal articles and conference papers) and *consolidated resources* (reviews, monographs etc.).

Table 3.5. Research stages according to Garvey

Phase	*No.*	*Stage*
Input	1	Problem perception / definition
	2	Initial formulation of solution
Throughput	3	Strategy for data collection
	4	Choosing technique for data collection
	5	Designing equipment / apparatus
	6	Choosing data analysis technique
Output	7	Relating data to known knowledge
	8	Interpreting collected data
	9	Integrating findings into science

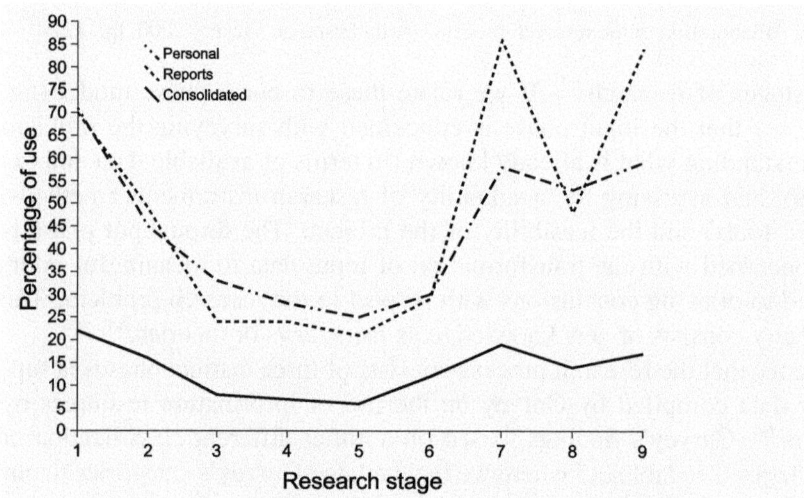

Data source: Vickery 2000, table 31, after Garvey 1979.

Figure 3.11. Information sources in research

certain period of time on the sequence of research activities, going through the input, throughput and output phases. Eventually this will result in publication of the research results, e.g. by means of a published research article. At that moment the research process enters into the public domain, adding new knowledge to the existing body of knowledge. The researcher(s) will then typically embark on a new research activity that will, after some period of time, create a new output. In the mean time, between the distinct outputs, the research activity remains more or less a 'black box,' hidden from the public domain and not contributing to the common body of knowledge.[65]

The discrete (non-continuous) nature of scientific reporting is predicated on the analog format that consists of distinct objects (i.e. documents). Will this be changed by the digitization of scientific communication? In theory this is possible if we look at certain properties of the digital format:[66]

- streaming media: the digital format can provide a continuous flow of information reflecting the 'current state of affairs' in real-time or at least by way of frequent (e.g. daily) 'snapshots'.

- networked access: the digital format can provide real-time access over the network to the research environment (e.g. the laboratory), using digital 'eyes and ears' such as sensors, video-camera's and databases.

However, it is too easy to suggest that digitization might lead to a more continuous form of output, for instance in the context of e-science. The reason for this is that continuous output must necessarily lack a basic requirement of scientific communication, viz. certification of the output. Certification of scientific output requires (a) a discrete, identifiable and fixed entity that is to be certified and (b) a certain period of time during which certification is carried out. This is a characteristic that distinguishes science from other activities that can be certified on the basis of a standardized process. For instance, in health care quality control is based on protocols that prescribe in precise terms how medical procedures are to be performed. Compliance with the protocols is then sufficient to guarantee the required level of quality. Since it is an objective of science to develop new procedures and new outcomes, protocol-based quality control cannot be sufficient. Quality control of the research output therefore cannot be a component of the research process itself but has to be externalized, based on the output itself rather than on the input and throughput of the research process.

8. The transition from print to a digital information chain

8.1 The digital information chain

One can wonder whether, after a history of some three centuries, the properties of the scientific journal are sufficiently strong to maintain closure, or

[65] More or less, because there may be some interim reportings at conferences or through research memoranda. But in terms of *formal*, certified contributions our model holds.
[66] See also chapter 4, where we describe digital properties in more detail.

whether contextual factors are now leading to a re-opening of scientific communication. At this point we observe that the scientific journal appears to have adapted to contextual changes without losing its position as the primary form for scientific communication. For instance, the journal has rapidly migrated to the digital, networked environment. Within this environment there exist many competing forums (e.g. electronic forums, discussion lists, e-conferences, portals etc.), but there is no evidence that these have re-opened formal scientific communication in a substantial way, let alone resulted in a new closure. This is confirmed by Nentwich's recent and comprehensive study on 'cyberscience'[67] that presents numerous speculative visions and scenario's, but little substantial evidence that a new form is emerging for the mainstream of scientific communication.

We now look at a rather different but nevertheless related issue concerning the structure of the information chain, viz the way scientific communication is organized and how this changes as a consequence of digitization. Here too, a significant level of closure has taken place with respect to actors, functions and procedures. Although, as we have seen, there exist various models of the information chain (representing as many different ways of viewing and discussing scientific communication), each of these models seems to be fairly stable over time. They suggest that closure of the process of scientific communication occurred fairly early, and has remained stable for at least the preceding century.

However, some more recent models do suggest that structural changes in the information chain may be forthcoming, with digitization as the key enabling factor. This issue is apparent in an interesting model that was developed by Aitchison in 1988.[68] What this non-cyclical model shows is the potential of electronic formats to transform the structure of the scientific information chain by providing *parallel* channels that are capable of bypassing more traditional actors (fig. 3.12). For instance, Aitchison describes database producers and online hosts (at that moment the precursors of the current web-based actors) as alternatives for the functions provided by libraries, at the same time adding 'intermediaries' that might negotiate between online services and libraries on behalf of the user. But Aitchison also allows for the possibility of direct communication between authors and users, although at that point in time the means for doing so (DTP, e-mail, bulletin boards) were not yet generally available. Aitchison recognizes that the entire system of interconnecting actors and functions in the information chain should facilitate communication, but in practice may not always do so: 'everything between the author and the user can be thought of, in one's optimistic moments, as facilitating the transfer of information between the two, or in one's less sanguine moods, as standing in

[67] Nentwich 2003.
[68] Aitchison 1988.

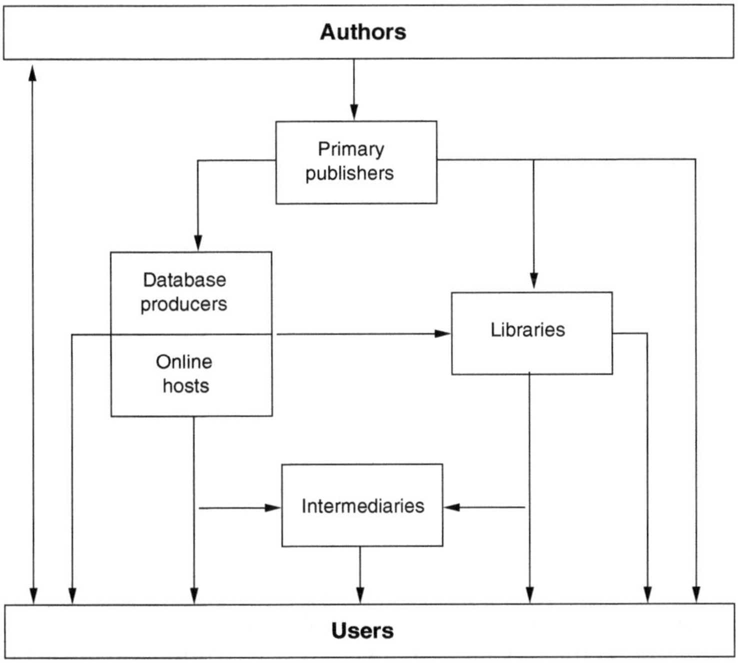

Figure 3.12. Aitchison's model of the information chain

the way of proper communication'.[69] Nevertheless, according to Aitchison, intermediary actors do have to provide added value through their functions, otherwise 'they would deservedly cease to exist'.

A far more radical representation has been suggested by Julie Hurd as a possible model for the future (fig. 3.13).[70] This model is a re-writing of the Garvey-Griffith model (see fig. 3.4) in terms of expected future developments, based on a number of assumptions:

- Research is considered to be carried out in collaboratories rather than by individual authors
- Research outcomes as well as underlying information are integrated in the research report
- All research-related outputs are in digital form

Hurd's model takes into account a number of phenomena that are already apparent in the (digital) information chain: the proliferation of pre-print servers

[69] Aitchison 1988, p. 320.
[70] Hurd 2000.

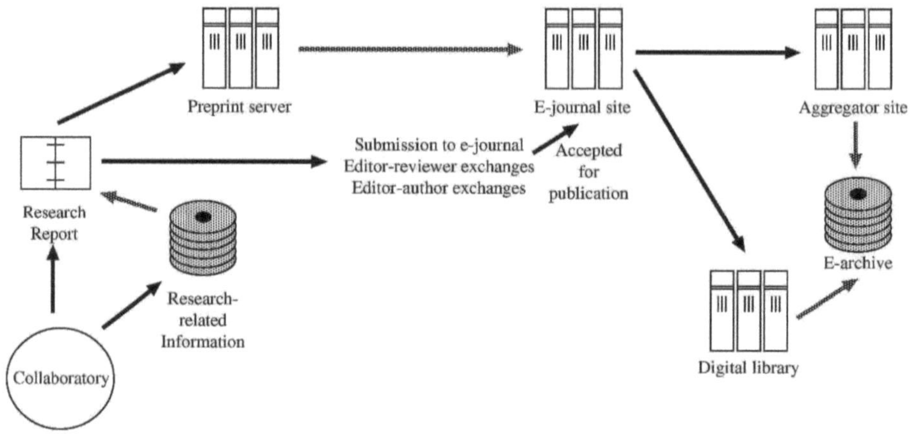

Figure 3.13. Hurd's model for 2020

(and other types of repositories), the increasing role of 'aggregators'[71] that serve as clearinghouses between journal publishers and digital libraries or end users, and the need for special e-archives for long-term storage. This points to an increase in the number of intermediary roles as a consequence of digitization. Another interesting feature of Hurd's model (although not mentioned explicitly by Hurd herself), is that it describes an increasing level of aggregation during the life-cycle of scientific articles (fig. 3.14). Working backwards from the end of the life-cycle, large-scale e-archives (e.g. on a national basis) will take care of the long-term storage of the collections of various digital libraries. Digital libraries already provide access to materials from different aggregators, which themselves incorporate larger numbers of e-journals. E-journals are based on materials that may already be distributed by different pre-print servers. Both E-journals and pre-print servers publish materials from larger numbers of collaboratories (or individual authors).

It is also interesting to note that the user (who is not explicitly included in Hurd's model) typically will use different access points depending on the stage in the life-cycle of the publication. Initially (and prior to certification) the pre-print server will serve as as the access-point. After certification and inclusion in an e-journal (i.e. typically after 6-12 months), access will shift to the digital

[71] Aggregators are companies that provide access (via their web-pages) on a subscription or license basis to groups of publications, on behalf of publishers and usually in combination with user-oriented services such as search and browse functions, cross linking etc. Aggregators perform in the digital world a similar function as subscription agents for print. Instead of selling subscriptions to hard copy, they provide links to electronic versions of journals that are produced by any number of different publishers.

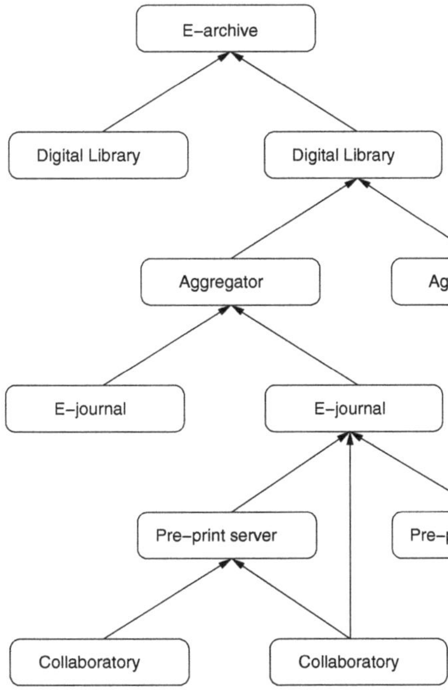

Figure 3.14. Aggregation in the digital information chain

library. Finally, when the publication loses its short-term value, access will shift to the long-term e-archive.[72]

This development towards *multiplicity of sources* is reflected in many models predicated on the digitization of scientific communication. Already in Aitchison's model, information is carried to the user by libraries, database producers, online hosts, intermediaries, and even directly by the author. Hurd's

[72]Long-term archiving of digital information is a new and separate function that has become necessary because of the limited longevity of the digital medium and the technical environment in which it operates. Attention to the problem of digital longevity was drawn by a number of studies published around 1996 (Task Force on Archiving of Digital Information 1996; Mackenzie Owen and van der Walle 1996). Since then, a number of research projects have concentrated on developing technical and organizational solutions. The main current solutions strategies are *emulation* (Rothenberg 1999; Bearman 1999; Granger 2000; Lorie 2000) and various forms of *migration* (Granger 2002; Holdsworth and Wheatley 2001; Wheatley 2001). The responsibility for the long-term archiving function is not yet clear. There are objections to a role for publishers, and it has been argued that national libraries should take on this role (Mackenzie Owen 1996). In the Netherlands, for instance, the Koninklijke Bibliotheek has set up a digital archive for scientific literature in cooperation with IBM (http://www.kb.nl/dnp/e-depot/e-depot-en.html) using a system developed in cooperation with IBM based on emulation and the concept of a 'Universal Virtual Machine' (UVC) (Rothenberg 2000, Lorie 2002.), and the US National Library of Medicine has created a digital archive through the PubMed Central repository (http://www.pubmedcentral.nih.gov/). There are also initiatives by scholarly associations, e.g. the PROLA archive of the American Physical Society that stores in digital form all journals published by the APS since 1893 (http://prola.aps.org/).

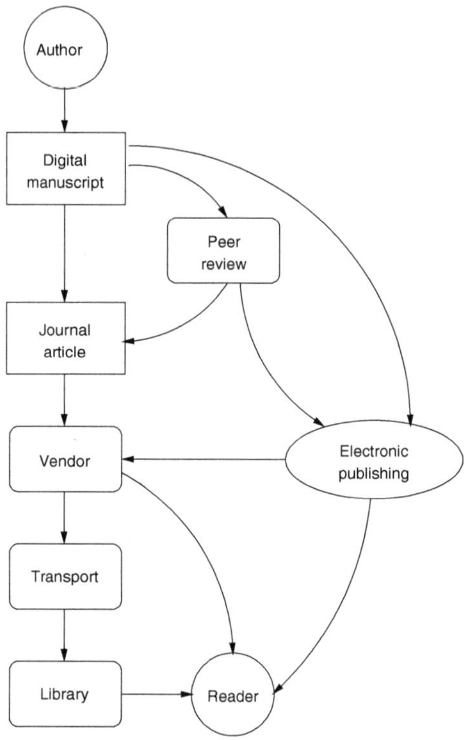

Figure 3.15. Fjälbrant's model of scientific communication

model depicts, as we have seen, various access points depending on the stage in the life-cycle of the publication. Also, journal submissions can flow to the journal directly from the author via the review process, or can be acquired from pre-print servers. A similar view is expressed by Fjällbrant (fig. 3.15).[73] Here various connections are envisaged between the author and the reader, involving refereed and unrefereed materials, formal channels and various networked forums.[74]

8.2 Systems based communication

The models described in the previous section are representations that attempt to describe and predict possible *developments* in the information chain under conditions of digitization. The basic notion underlying these future-oriented models is a transformation from an actor-mediated organization of

[73] Fjällbrant 1997.
[74] What Fjällbrant describes as 'electronic publishing' should be interpreted as a variety of self-publishing modes on the Internet.

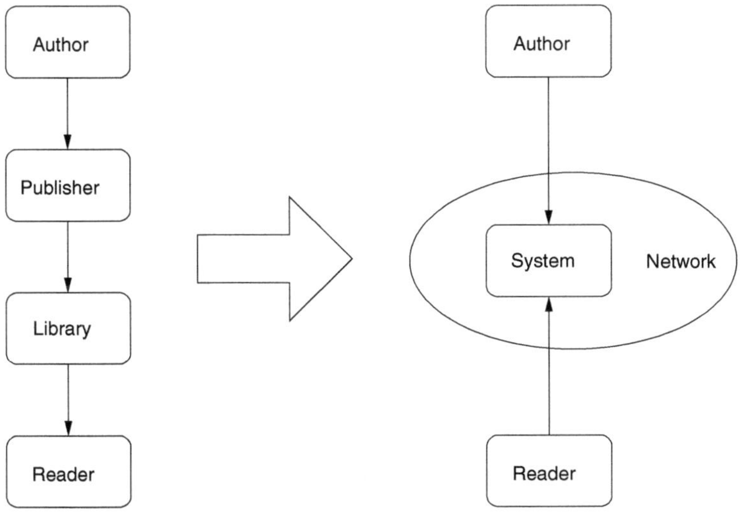

Figure 3.16. Transformation of the information chain

scientific communication to a system-mediated approach that is felt to be made possible by the digitization of scientific communication formats (fig. 3.16). Such a transformation would imply that many functions of the information chain can be carried out more or less automatically (and therefore become transparent and embedded in the communication system), rather than being performed by human and institutional agents. Many observers believe that the current system has a number of negative properties (such as biased peer review, power imbalances, excessive cost, and in general terms the dependency on extra-academic, commercial actors), and that a transformation towards a more system-mediated model might be capable of removing the imperfections of the current system.

However, as Hurd's model makes clear in a subtle way, removing human intermediary actors from the equation is not that easy. Although her model suggests a systems-based sequence of processes, it also includes editor-reviewer and editor-author exchanges, in other words: institutionalized peer review as a condition for certification and acceptance.[75] A similar approach to Hurd's is found in the 'Scholars Forum' model proposed by Buck, Flagan and Coles as a way to 'redefine the paradigm for scholarly communication'.[76] Here too, peer review as well as other functions such as copyright management stand in the way of a totally automated flow of information. In fact, if we accept that

[75]As Hurd writes: 'I believe that peer review will be a feature of any new communication system, although the mechanisms to ensure quality may be different in a digital submission and review process' (p. 1281).
[76]Buck et al., 1999.

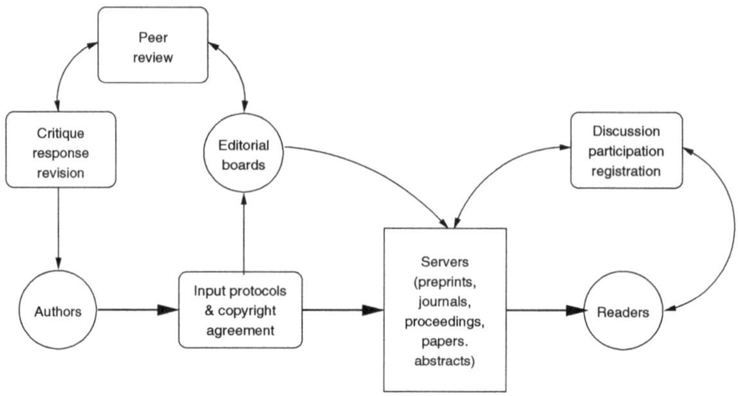

After Buck et al., 1999.

Figure 3.17. The Scholar's Forum model

the functions of intermediary actors (e.g. publishers and libraries) as described in table 3.4 (p. 65) are relevant and necessary, they would have to be provided by any systems-based solution for scientific communication. It is difficult to imagine how most of such functions could be performed automatically, without the intervention of institutionalized actors.

However, this does not mean to say that the actor-based organization of the information chain will remain stable despite digitization of scientific communication formats and the proliferation of networked communication. To explore the effects of the transition from print to digital publishing, we have developed a distribution model that focuses on two basic functions of the information chain: selection and storage (described in the model as a 'memory' function).[77] Fig. 3.18 describes the traditional system of print distribution, where the memory function is institutionalized as a repository within an information centre (e.g. as the collection in a library). A first selection point is found in the interaction between author and publisher (usually mediated through the editorial board and peer reviewers). The information centre contains two selection points: one where the centre selects materials to be acquired and added to the collection, and another where the user selects relevant information from the collection. The flow of information from author to user is negotiated at these three selection points. As in our model mentioned above (fig. 3.5 and 3.6), the information chain is represented here as a transaction space. Using this form of representation, fig. 3.19 shows a set of alternatives for a digitized information chain, showing various (theoretical) options for embedding selection and storage:

[77]Mackenzie Owen 1998. This approach builds on the concept of selection inherent in an earlier model (from Mackenzie Owen and van Halm 1989) discussed on page 57.

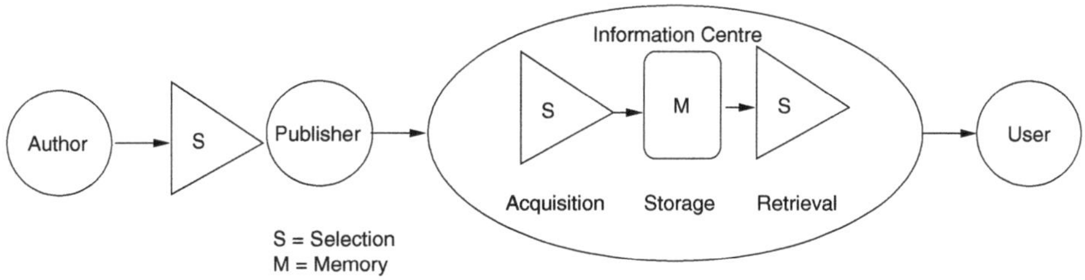

Figure 3.18. Print distribution

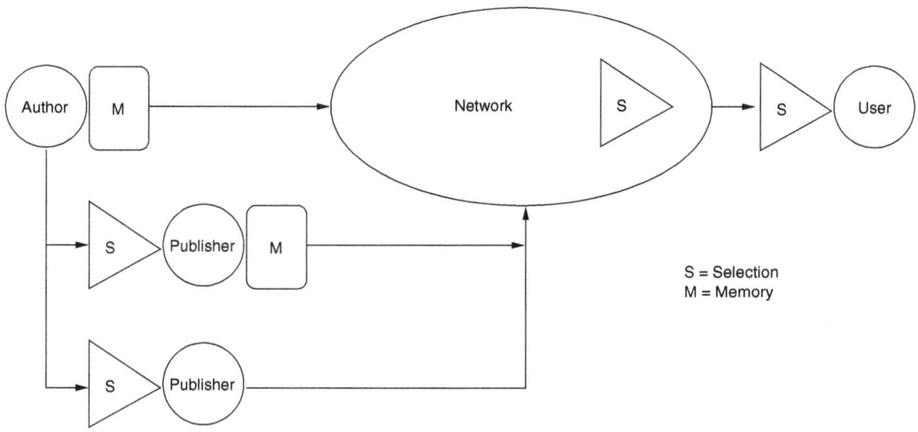

Figure 3.19. Digital distribution

1. The most radical option is the elimination of intermediary actors: users select over the network materials that are held in store by the author;

2. A different solution is to eliminate only the role of the information centre and to create a repository managed by publishers; in this case users access materials over the network that are held in store by the publisher;

3. A final solution is to eliminate the memory function altogether and to adopt a 'push'-model: publications are sent directly over the network to users, based on personalized selection criteria (profiles).

These models demonstrate that even if the memory function (repository) is eliminated, the information chain remains a transaction space because the selection function allows the user to accept or reject what is offered over the network and to choose the sources of profile-based selections.

8.3 Institutional repositories

The traditional information chain, as represented by a number of models described in the previous sections, is a monolithic system based on fixed actors with specific roles and functions. What we are beginning to see now is a far more differentiated approach, based on functionally defined 'building blocks' that are available to authors (and users), and from which a specific trajectory can be constructed through which information flows. This approach reflects the vastly increased number of options for creating, storing and disseminating information caused by digitization and the proliferation of networks. A key issue is the fact that this development allows multiple uses for the same information (e.g. as a journal publication, a component of a web site, and as an entity within a digital teaching environment).

Representations that challenge the traditional, actor-based information chain reflect a desire for alternative distribution channels as well as for more control over the research output created within an institution or disciplinary domain. A solution may be found in the concept of *institutional repositories* that provide a standardized and functionally neutral environment for storing and maintaining control over scientific output, and for making information available for a variety of purposes and types of users.[78] This approach is exemplified in what we call the Data-Services model (fig. 3.20).[79] In this model a repository is used as a clearinghouse, servicing a number of (digital) forums for the dissemination of scientific information.[80] In terms of formal scientific publication, the e-journal is embedded in a range of communication forums and the peer-reviewed article emerges as but one of a multiple set of views on the information contributed by the author(s) to the repository.

A conceptually different way of representing this approach is by means of a 3-layer model consisting of:

1. a work space that provides the research or research group with a virtual work environment including tools for data acquisition and manipulation, and intermediary storage;

2. a storage space (repository) where the output of individual work spaces are aggregated and managed;

[78]Crow 2002; Lynch 2003; Shearer 2003. Institutional repositories are generally based on standards and protocols developed by the Open Archives Initiative (http://www.openarchives.org/).
[79]Mackenzie Owen 2003.
[80]Note that institutional repositories can also be designed to include additional materials (e.g. datasets, teaching materials etc.), or in general any information object created by a member of the institution. An alternative is the disciplinary or domain-based repository.

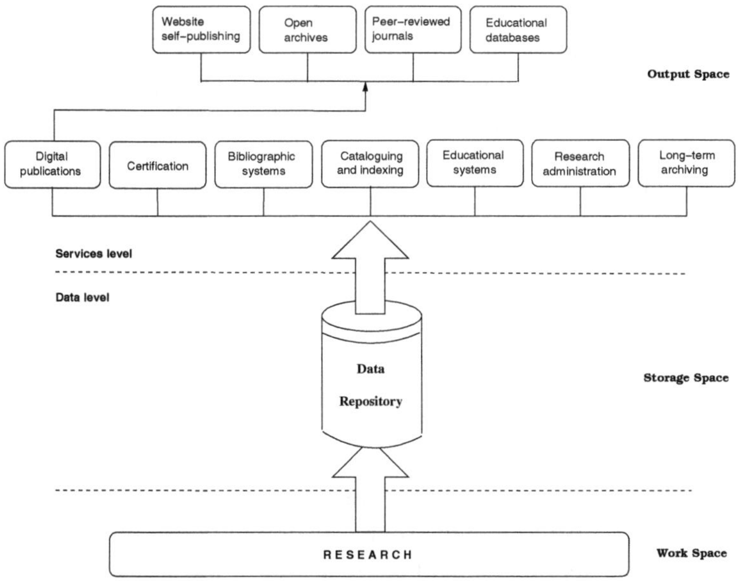

Figure 3.20. Data-Services Repository Model

3. an output space where various functions (services) are applied to information in the storage space, either to provide direct access or data harvesting, or for creating a variety of information products for dissemination.

A further elaboration of the repository-based approach has been described in a recent paper by Van de Sompel et al., (2004). They argue that the established scholarly communication system is dominated by the journal article as the unit of communication. They object to this state of affairs for a number of reasons:

- the current communication system does not sufficiently support a research practice that has become highly collaborative, network-based and data-intensive;
- the scholarly communication system is not able to cope with other units of communication, such as data, simulations, informal results, preprints, etc.;
- non-textual materials (e.g. multimedia) are at present treated as add-ons rather than essential parts of the publication; they cannot easily be incorporated into the existing publication model or treated as publications in their own right;
- the integration of peer review in the publication process causes significant delays in the dissemination of scientific information.

Van de Sompel et al., propose a re-design of the scientific communication system based on a wider range of knowledge representation types (including compound documents aggregating the basic types) and on *registration* (e.g. when first deposited in a repository) rather than *certification* as the key function that allows information to be made publicly available. This re-designed system would be based on a functional model, incorporating a number of functions distinguished by Roosendaal and Geurts (1997) that we have already mentioned: registration, certification, awareness, archiving and rewarding. The ideas is that (flexible combinations of) these functions could be implemented by multiple actors in different ways. The information chain would then consist of multiple trajectories through the various functions as provided by different actors. The trajectory followed by a specific publication could depend on the type of communication unit, the scientific domain, the available infrastructure, and even personal preference.

Van de Sompel et al., believe their approach to allow for more innovation, adaptability and competition in the scholarly communication process. In more political terms, this approach might cause a shift of the locus of control in scientific communication from the publishing industry to academic communities.

As with the general repository model described above, this approach is based on a loosening of the traditional, fixed relationship between actors and functions in the information chain. However, whereas the general repository model is now slowly emerging through the development of institutional repositories in many domains, the approach of Van de Sompel et al., would require a significant level of 'social engineering':

> '... our proposals are mainly technical and architectural, but with wide ranging social and organizational implications. Like any technology, success will depend not only on technical soundness but on the willingness of the participants in the system – publishers, scholars, academic institutions, funding institutions, and others – to adopt new tools and develop new organizational models on top of them.'[81]

In a more formal sense we can describe repository-based scientific communication as a function of research output (as stored in a repository) and two operations performed on the stored output: research-related certification (e.g. some form of peer review) and user-related aggregate functions (e.g. profile-based alerting, cross-linking, search functions, rights management etc.). We call the latter functions 'aggregate' because they are performed at the aggregate level on entire collections of documents rather than (as is the case with certification) on individual documents. This representation focuses on the three key functions of the communication system (storage, certification and aggregation) without relating these functions to specific institutional actors (fig. 3.21). It reflects the current state of affairs where there is no clear relationship any more

[81] Van de Sompel et al., (2004, Conclusion).

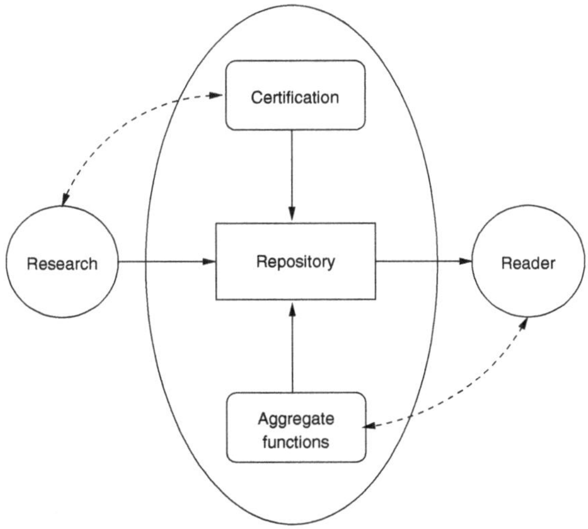

Figure 3.21. Repository-based communication

between actors and functions, and some functions may become system properties embedded in the network rather than remaining based on institutional roles.

9. Innovation of the scientific journal: a socio-technical interaction model

Models such as those developed by Lancaster, Hurd and Fjällbrant, as well as the life-cycle model developed by Björk and Hedlund limit their representation of the information chain primarily to the various stages involved in publishing research outcomes. They abstract from the social context by representing scientific communication as an isolated, autonomous process that is based on formalized genres such as the research article and the conference paper. In reality, scientific communication is a more complex phenomenon, involving a large array of genres, technical environments and social contexts. The UNISIST model mentioned above (p. 3) was a first attempt at identifying the various genres involved. Kling, McKim and King have done significant work on expanding the notion of scientific communication beyond formal publication and on placing it in the context of social reality.[82] This is necessary because scientific communication is – as we have seen represented in more recent models – no longer based on a stable, monolithic system, but increasingly on

[82]Kling 1999; Kling and McKim 2000; Kling et al., 2003.

a dynamic, multi-faceted system where new (and often digital) modes of communication are developed and are competing for a role in the scientific arena.

9.1 Scholarly Communication Forums

Kling, McKim and King have introduced the concept of *Scholarly Communication Forums* (SCFs) as a neutral descriptor for any type of communication channel, irrespective of its technical characteristics, genre or communicative role. A specific class of SCFs is the electronic SCF (e-SCF), defined as an SCF based on computer-mediated communication. It would be relatively easy to extend traditional models of scientific communication to include these new forums. Electronic SCFs are, however, not just technological alternatives for established, non-electronic SCFs. For instance, communication forums based on new digital genres also allow the emergence of relationships between domain experts in the form of specific (and often relatively closed) *networks of practice*: 'Electronic ties are loosening the constraints of organizational structure and physical proximity to allow connectivity between individuals who would otherwise find it difficult to identify and sustain contact with others who share the similar interests'.[83] In other words: digital genres are capable of engendering new social networks for the exchange of information.

An interesting question is why people choose to use new forums for communication. As outlined in the previous chapter, technological deterministic explanations are too simplistic because they neglect the social and psychological factors involved in the development and application of technology. It has been argued for instance that the emergence of new digital forums for scientific communication is not only predicated on new technologies, but also on the control structure of established forums. New forums would then provide alternative channels for those denied access to established forums.[84] Kling, McKim and King note that the common approach in the literature is to see the adopters of new communication technologies as motivated by the technological features of alternative forums in terms of perceived benefits (or drawbacks), and also as being free to adopt a specific e-SCF or not. We believe that this approach is reflected in and reinforced by representations (such as various models described in this chapter) that focus on communication as a technical process and that abstract from other contextual factors.

Kling, McKim and King adopt a different approach that is based on the concept of Social Construction of Technology (cf. 1.1 on page 191) and on Latour's Actor-Network Theory.[85] Emphasizing the interaction between technology on the one hand, and the form, purpose and effects of communication

[83]Faraj and Wasko 2001.
[84]Sullivan 2000.
[85]Latour 1987.

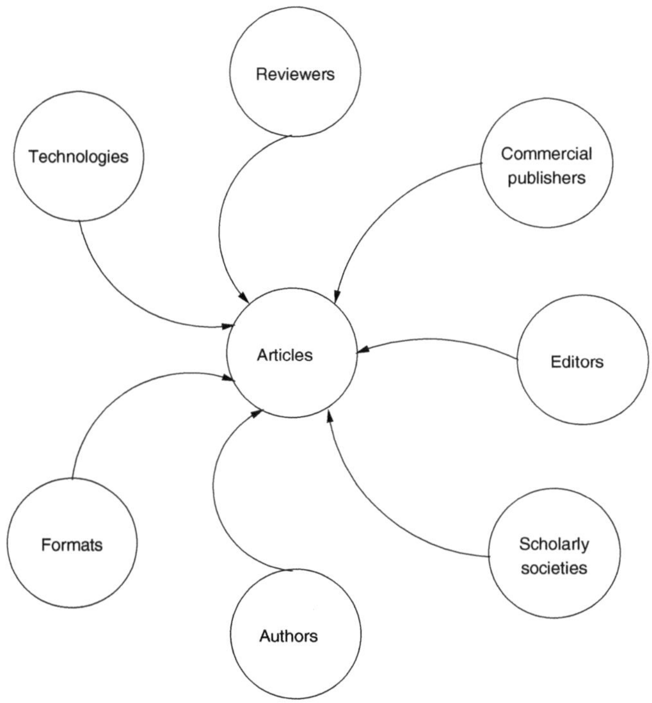

Figure 3.22. STIN model of scientific journal production (After Kling et al., 2003)

forums on the other, they describe scientific communication forums in terms of Socio-Technical Interaction Networks (STIN) of 'system interactors.' For instance, a scientific journal can be represented as a network in which information materials, technological applications and human actors form a specific configuration (fig. 3.22).[86] This configuration itself is embedded in yet another configuration: a socio-technical network encompassing authors, readers, institutions (e.g. libraries, scientific societies, editorial boards, academic committees) as well as technological artefacts, rules and procedures (fig. 3.23).

9.2 The innovation chain

How does innovation work within the scientific communication system, seen from the viewpoint of Socio-Technical Interaction Networks? We begin by representing the various factors involved in the innovation of the scientific journal within a socio-technical network of the electronic journal/article (fig. 3.24). In this network we distinguish the following system interactors and their interrelationships:

[86]It should be noted that there also exist manifold network relationships between authors, editors and reviewers, with scholarly societies and commercial publishers, and possibly also with developers and implementors of technologies and formats.

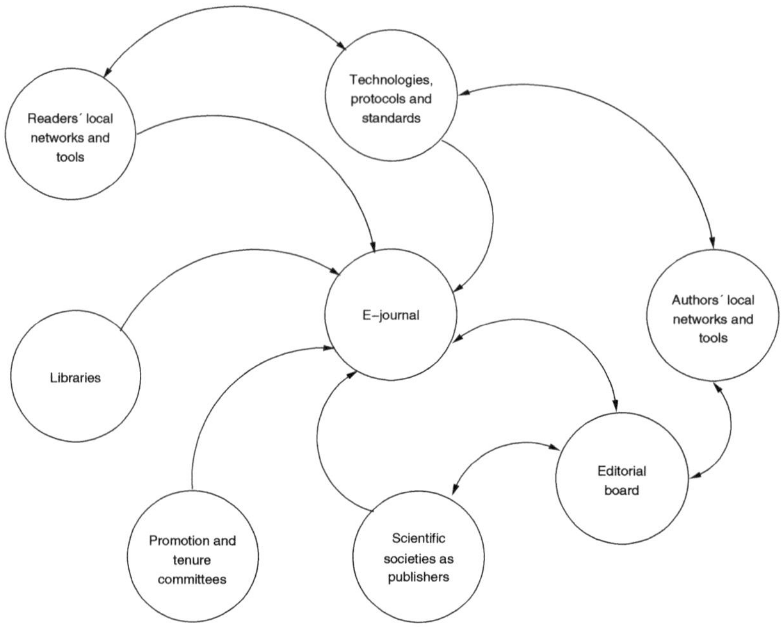

Figure 3.23. STIN model of an electronic journal (After Kling et al., 2003)

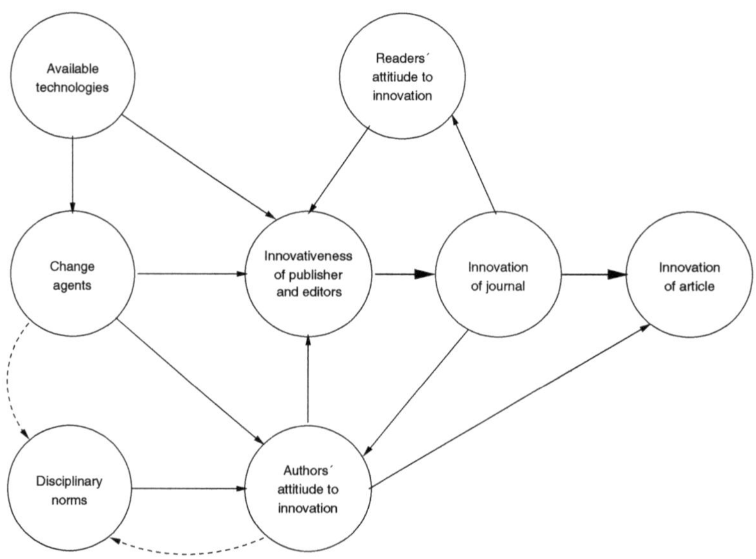

Figure 3.24. A socio-technical network of journal innovation

- The level of innovation found in individual articles is governed by conditions set by the journal in which it is published. For instance, if the journal does not allow moving images, they will not be included in articles. Note that this is only a conditional factor: if the journal does allow a certain innovative property, it will not necessarily be used by authors in their articles. The level of innovation of the article is therefore determined by the degree in which authors use the innovative properties available within the constraints set by the journal.

- The level of innovation of a journal depends on the innovativeness of the publisher who has to provide the technical context for innovative properties, and on the editorial board in accepting the innovative properties made available by the publisher.[87] The journal's innovativeness is a co-determinant of the author's decision to publish in that journal.

- The innovativeness of the publisher is, of course, a highly complex issue, governed by many factors including the publisher's long-term commercial strategy. Here, we limit our analysis to the following factors:

  - available technologies: these are the technologies from which the publisher can choose to use for innovating the journal;
  - change agents that put pressure on publishers to take a certain course of action, e.g. to adopt innovative technologies;
  - authors for whom the innovativeness of the journal is a factor in deciding to publish in that journal; positive or negative reactions of authors to innovative properties of the journal will have a positive or negative effect on the innovativeness of the publisher;
  - readers, whose attitude towards innovation is influenced by examples in the journal literature.

- The attitude of authors towards innovation; this is determined by at least three factors:

  - their (positive or negative) response to examples of innovation in the journal literature;
  - the influence of change agents (notably peers);
  - the norms prevalent within the discipline.

- Disciplinary norms; these tend to have a conservative influence on the innovativeness of actors within the social group. However, in the long run,

[87] Note that we distinguish between (innovative) properties of the journal as an aggregate form (communication forum), and (innovative) properties of the article (genre).

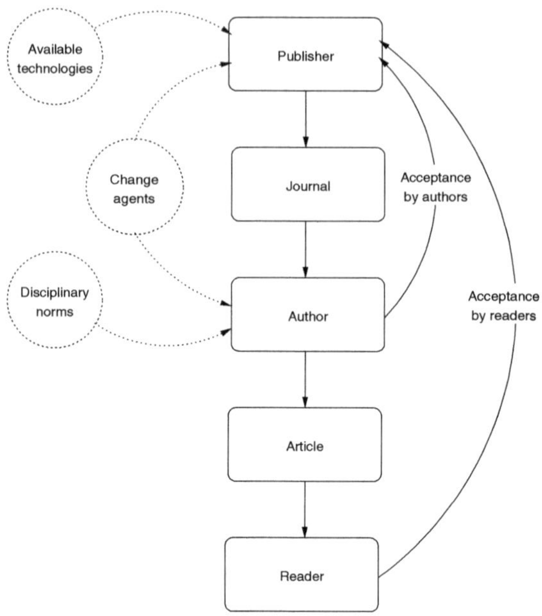

Figure 3.25. The innovation chain

these actors themselves (including change agents) bring about changes in the disciplinary norms.

Using this analysis, we now re-draw the network representation of fig. 3.24 to derive a model that represents the flow of innovation over time through the scientific communication system as an 'innovation chain' (fig. 3.25). Here we see that it is ultimately the *publisher*[88] who – influenced by technological options and change agents – initiates innovation by introducing new potential properties in scientific journals. These innovations are then taken up (or rejected) by authors in two ways:

- authors react (positively or negatively) to changes at the aggregate level of the journal; these reactions influence the publisher's willingness to continue with the innovation; at this point the innovation is reflected in the communication forum, i.e. the scientific *journal*;
- authors accept or reject potential innovations at the specific level of the article, by incorporating (or not) new properties in their writings; it is only at this point that the innovation is reflected in the genre of the scientific *article*.

[88]Note that 'publisher' refers to any personal or institutional actor performing the publishing functions listed in table 3.4, i.e. not necessarily an 'official' or 'commercial' publisher.

Finally, as innovations incorporated in scientific articles and/or embedded in journals reach the reader, they are subjected to a final stage of acceptance or rejection. If readers approve, publishers will continue the innovation process, otherwise the innovation will be discontinued.

10. The complexity of scientific communication

In this chapter we have analyzed the scientific communication system in terms of a large number of models that represent different views of its structure and of its development towards digitization. Most of these models are essentially based on the conduit metaphor, represented in its basic form by the concept of an 'information chain', and describing a flow of discrete information objects from author to reader, mediated by a number of intermediary actors performing specific functions. In section 7 we have looked in more detail at the role of the author as an actor in the information chain and at the relationship between stages in the research process and scientific communication. The transition from print to digital communication is reflected in a number of models that we have analyzed in section 8.

In section 8.3 we have presented a two-tiered data/services model of the emerging repository-based communication mode that shows a break away from the former, fixed communication system towards a system that is more flexible and is 'constructed' according to contingent needs and opportunities. Finally, in section 9, we have taken up the issue of innovation again, and looked at the socio-technical context of scientific communication. Here we developed a model of the innovation process as it evolves through the communication system.

One conclusion that emerges from the various models discussed in this chapter is that the conduit model is an inappropriate representation of the way scientific communication operates. Our analysis shows that the information chain functions more as a transaction space with a number of selection points where various actors negotiate inclusion in and acquisition from the certified body of scientific knowledge. The concept of a transaction space encompasses the idea of engagement, the active involvement of the user in determining the flow of information. As a consequence, scientific communication has to be seen within a constructivist paradigm where there is no single, objective 'scientific state of affairs' but a multitude of 'world views' informed by the specific engagement of the individual scientist with the scientific communication system.

Another shortcoming of the conduit model is that it simplifies the role of the author without taking into account the complex relationships between research practice and information. We have therefore described the author's role in terms of a 3-phase model that distinguishes between the input, throughput and output phases of research. We have argued that the input to research consist of both data created by the research activity itself, and external data and documents reflecting the prevailing body of knowledge. The type and role of external information as well as the extent to which it is used, are different in the various stages of research. In addition, scientists are engaged in an ongoing communication activity with peers, outside the formal information chain. Nevertheless, formal scientific communication remains a discrete process where scientists output research results periodically rather than continuously. Although digitization in the context of 'e-science' could lead to a more continuous form of communication, the requirement for output-based rather than process-based certification remains a problem.

An important issue in discussions of the scientific communication system is its functionality in relation to the various actors. The traditional communication system has evolved towards a high level of closure with respect to functionality, actors and role divisions. However, it seems that digitization is opening up the system considerably. Publishers and libraries are taking over each other's functions, some functions are becoming embedded in the digitized network system, and others are being taken over to a certain extent by authors and readers or their parent institutions (e.g. in the form of 'self-publishing'). In addition, various new functions have emerged (such as pre-publishing and long-term archiving) that require some form of control and actor-involvement. Ideas have been developed for a more radical effect of digitization, leading to a transformation towards an entirely systems-mediated form of communication without any involvement by institutional actors. We have seen, however, that many important functions of the information chain, and especially the certification function, do require such involvement.

In fact, rather than becoming more simple (cf. fig. 3.16), scientific communication is becoming more complex. The flow of scientific information from author to reader is no longer a single, well-defined process, passing distinct stages governed by actors with strictly assigned roles. Digitization has resulted in an increase in genres, actors and communication modes, with multiple trajectories through which scientific information can flow, and multiple access points for users to acquire that information, depending on the stage in the life cycle of the publication. This complexity increases even more if we consider that life cycle models such as that developed under the SciX project follow the traditional notion of the conduit model where the scientific publication changes perhaps as to form, but not as to content as it traverses the information chain. However, it is also possible to regard scientific communication as a process in which the content and significance of scientific output evolves over time through the information chain. This process includes at least the following stages (fig. 3.26):[89]

[89]Further expansions of this model can be made, for instance by adding a *citation phase*, i.e. the phase in which a document is no longer an isolated entity, but becomes embedded in the citation network of science.

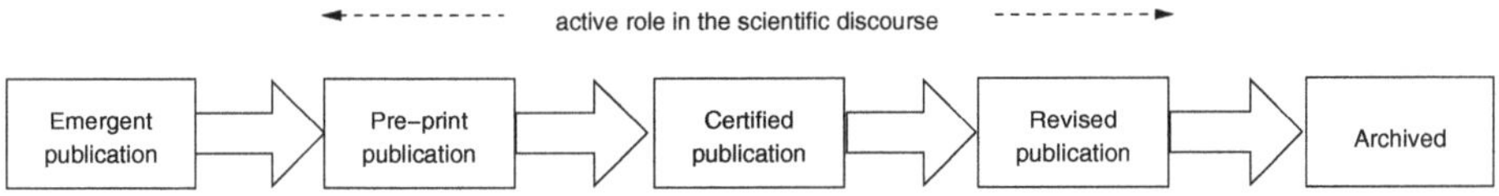

Figure 3.26. Continuum of scientific publications

- *emergent phase*: in this phase research results come out of the research process and are transformed into some form of information that can be recorded and made public
- *pre-certification phase*: in this phase intended publications for which certification (e.g. peer review) has not been concluded are made publicly available via repositories and pre-print servers; the document begins its active role in the scientific discourse
- *certified publication phase:* here the publication is made available as a certified (peer reviewed) document
- *revision phase:* in this phase the certified document is revised (e.g. annotated or expanded) due to apparent shortcomings, new insights etc.[90]
- *archived publication phase:* at this stage the publication no longer plays an active role in the research process (and is possibly no longer valid in terms of its findings), but remains as a record of the scientific heritage.

In other words, it seems to be increasingly difficult to consider the scientific publication as a fixed, well-defined entity. The formal, certified scientific publication that is the primary object of this study, emerges as but one specific (though central) manifestation of scientific output that may be preceded and followed by other manifestations.

The various models described in this chapter are as many representations of how information science perceives the domain of scientific communication. It is interesting to see how perceptions have evolved over the years. Initially, models were developed in order to understand and come to terms with the way scientific communication operates. The UNISIST, Lancaster and Garvey-Griffith models are examples of this approach. At a later stage models were developed in order to come to terms with developments – especially related to digitization – in the information chain, and to predict the outcomes of these developments. Examples of this are models proposed by Aitchison, Fjällbrant

[90]The possibility of revising a publication is often regarded as an (either advantageous or dangerous) property of digital publishing formats. Note that this concept is somewhat contentious and may not be acceptable in many scientific domains. See also chapter 4.

and Mackenzie Owen. Finally, recent models such as proposed by Buck et al. and Van de Sompel et al. informed by an engineering paradigm in that they present ideas about how scientific communication *should* be re-designed in order to resolve certain problems or to meet specific goals.

If we observe the various modes of thinking about scientific communication as they have evolved over the past fifty years, we also see a shift from a linear, technocratic and systems-based view to a view based on a conception of scientific communication as a social system. The scientific communication system consists of a complex web of relations within a socio-technical network. This view does not only allow for a much richer analysis of the information chain, but also explains – as already outlined in the previous chapter – that innovation is not a straightforward technological process. We have concluded our analysis by constructing a model of the innovation process of the scientific journal. This model shows how various components of the information chain interact and are influenced by a number of external factors, notably the available technologies, change agents and disciplinary norms (fig. 3.25). Our final conclusion based on this innovation model is that the publishing function is the key component governing the innovation of the scientific journal in terms of its digitization. It comes therefore as no surprise that the various actors in the information chain – the academic community, libraries and new entrants as well as traditional publishers – are seeking to obtain control over the digital publishing function. As a consequence the structure of scientific communication has entered a state of instability. Closure towards a new structure of scientific communication in the digital environment will depend to a large degree on the outcome of this process.

Chapter 4

THE DIGITIZATION OF INFORMATION RESOURCES

1. Introduction

In order to study the impact of digitization on the scientific article, we need to obtain an understanding of what digitization is about and of what it means, at least in principal, for a document to be 'digital' rather than 'non-digital'. What happens to information when it is recorded and distributed in digital form rather than in written or printed form on paper? Is there any significant difference other than just a change of physical medium? Or does the 'digital' affect structure and content as well? And if so, to what extent does this impinge on scientific communication? What are the possible consequences of digitization for the scientific article and the way authors use the scientific article to communicate the findings of their research? These are the questions we shall address in this chapter.

We begin by considering the nature of the concept of 'digitization' itself, especially in relation to the document. We then look at the transition from analogue to digital in order to see how the properties of the document might change when it becomes digital. We shall discuss a number of general aspects of digitization that relate to fundamental differences between the analogue and the digital. These aspects include the *networked* nature as well as the *infixity* and dynamic characteristics of the digital document. These characteristics of digital information have consequences for two other issues that we shall discuss: the concept of *authenticity* and the level of *control* over form and content that is exercised by actors in the information chain. We also introduce the concept of *functionality* as a fundamental characteristic of the digital document. Most aspects of digitization discussed here pertain to documents in general. In the final section of this chapter we therefore look more specifically at digitization of the scientific article.

2. The concept of digitization

There is a striking difference between the nature of the information processed by computers in the early days of information technology, and the information that most people access and process with their computers nowadays. Initially, the application of information technology was aimed at 'data processing' and 'number crunching': using the computer for storing and processing large numbers of numerical data, either acquired by data entry (e.g. in administrative processes) or through sensors (e.g. in geology or space research). Numerical data processing is still of major importance today. Not only scientific research, but also many aspects of daily life (e.g. banking, power supply, traffic control, aviation, public administration) rely on the data processing power of computers, and the use of networks for distributing and exchanging data.

An important difference between the early stages of computing and present-day information technology is to be found in something that was lacking in the initial stage, and has now become ubiquitous: textual and graphical information resources that convey meaning to human beings, playing a major role in their exchange of knowledge, insights, opinions and ideas. The application of information technology is now predominantly aimed at creating, distributing, storing and accessing products of the mind, and at communication rather than at processing. This development is, of course, reflected in the language used: the shift from 'computer' and 'automation' to 'information and communications technology' mirrors the shift from data processing to the exchange of meaningful messages between humans. For most people, that is what using the computer and the network is about. Data processing and 'computing' in its original sense have been relegated to the specialist and private realm of data centres and supercomputing.

At the heart of this development lies the transformation of information-resources from analogue to digital formats that we shall describe here as a process of *digitization*.[1] The term *digitization* is problematic in that it is used in many different ways, varying from concrete and specific to broad and abstract. On the one hand the term is often used to denote simply the conversion from analog to digital information of a specific information object or collection of objects,[2] e.g. digitizing a book or library collection. We find this use, for instance, in numerous 'digitization projects' undertaken by libraries and other institutions, with the objective to make analogue collections accessible over the Internet, or to preserve in digital form materials that in their original paper

[1] The term digit*al*ization is also used (e.g. by Briggs and Burke 2002). We prefer the shorter and by far more commonly used term.

[2] This is the definition commonly adopted in dictionaries. It implies basically the symbolic representation of analogue data on a recording medium through an encoding of textual and graphical symbols onto a binary coding scheme.

form are threatened by physical destruction. At the other side of the spectrum we find uses of the term related to fundamental notions of modernity and the development of a globalized, network-based society. Used in this way, the notion of 'digitization' is an historical concept that offsets the current state of the world against the pre-computer era. For instance, Gunn notes that:

> 'to address the societal consequences of digitalization, the first order of business is to give a new meaning to the definition of society. ... In this paper, we will observe new digital societies coalescing around access to and utilization of personal communications, computational resources and electronic consumer goods, all of which are born out of the digital revolution.
> The most notable impact of the digital society is that it empowers the individual. This empowerment emphatically and undeniably changes his or her relationship to and dependence on existing social structures – personally, locally, regionally, nationally and, for the first time ever, globally'.[3]

Nwaobi (2001, ch. 5) describes, in a rather Utopian ecological context, a vision of a new society in which humans live in harmony with each other and with nature based on a transition that is described as digitization driven by knowledge and by the technologies for processing and communicating it.

The term 'digitization' therefore can denote both the process of transformation and the underlying causality and the role of the 'digital'.[4] The emergence of the Internet as the embodiment of digitization is related to a 'transformation of modern societies' – more specifically: the 'transformation of the spatial and temporal organization of social life', where the Internet is conceptualized as a modality of cultural transmission.[5] An extension of this approach leads to the notion that digitization is not about specific objects nor about society at large, but about the human being in modern society:

> '... Thus, the key cultural consideration of the Internet is not so much the digitization of information, but the digitization of us'.[6]

What this in fact means is that digitization and the transformation from analogue to digital can also be described in behavioural rather than technical terms. We (usually) do not create information in analogue form to subsequently transfer it to digital form. Information is created directly in digital form, and in this sense the process of digitization is a behavioural transformation: the digital mode has become the common mode of expression, documentation and communication. As far as the *writing process* is concerned, the digital mode is in

[3]Gunn 2000, p. 3.
[4]Another, very different meaning is given to digitization by Dretske, who uses the term as a descriptive model to describe how perception is transformed into cognition (Dretske 1981; Bonnevie 2001). Perception is seen by Dretske as 'analogue' (unselective, vague, imprecise, un-conceptualized), while cognition is similar to the 'digital' (selective, clearly defined, precise). Conversion to the digital happens for instance when we regard a landscape with its myriad of details, but only 'see' and allow to enter into our cognitive domain a single detail (e.g. a bird in the distance).
[5]Slevin 2000, p. 5-7.
[6]Federman 2003.

many respects different from its analogue counterpart.[7] Ferris even argues that digital (or 'electronic' as she calls it) writing is not writing at all, but a mediated form of oral discourse. Nentwich has argued in the context of scientific communication that the properties of the digital format influence not only the way a scientist gives expression to facts, ideas, arguments and findings in writing the text, but also influence the substance of research itself, i.e. the choice of research topics, of methodology, data processing etc.[8]

The more specific approaches to digitization focus on procedures, on 'how to' move from one type of medium to another. The broader, societal approaches focus on effects and outcomes of digitization on social and cultural systems, taking the digital format and its embodiments more or less for granted. In this section we take yet another approach by considering digitization of *the document* as a medium *for (scientific) communication.* We shall clarify how the document as a carrier of information changes through digitization, i.e. the consequences for the document of a shift from the analogue to a digital format. We are interested in how the properties of the digital document differ from those of its non-digital counterpart and of what *happens to the document* through the transformation from analogue to digital. By detailing the consequences of digitization for the document in general, we obtain a reference against which to measure the impact of digitization on the scientific article as a specific genre.

It is sometimes argued that digital media are similar to analogue media, except for certain technical characteristics that are usually described as 'added values' with reference to traditional media, e.g. storage capacity, speed of access etc. The traditional, analogue document is in fact capable of being transformed into digital form in a way that does not necessarily alter the original in any fundamental sense.[9] This happens in the case of digitization of the carrier, the 'document as thing' rather than of the 'document as knowledge'.[10] In many areas this basic form of digitization is being materialized, and in some areas the analogue form is in effect almost entirely being substituted for by the digital artifact.[11] This applies both to the physical document as, in an abstract sense, to the document as genre. This is notably the case within the domain of scholarly communication. It is estimated, for instance, that over 95% of all STM-journals[12] are now available in digital form.

[7]Dicks and Mason 1998; Bolter 2001; Ferris 2002.

[8]Nentwich 2003, p. 453–456.

[9]This can be proved by the fact that the digital version can be transformed back to an analogue form that is indistinguishable from the original.

[10]SeeBuckland 1991a,b.

[11]To give some statistics that underline the extent to which the digital has taken over from the analogue (from Charles et al., 2003): 92% of all information created worldwide in 2002 was stored on magnetic media; the World Wide Web is in volume seventeen times the Library of Congress print collections; e-mail annually generates a volume of information that is about 4.000 times the print volume of the LC.

[12]STM stands for Scientific, Technical and Medical.

It should be noted that a shift from analogue to digital *media* is not precisely the same as a shift from analogue to digital *writing*. Most writing is done nowadays with digital means, i.e. using computer-based tools for word processing, drawing and design, visualization, data management etc. Most analogue information (i.e. printed on paper) is created as the output of the digital writing process, and much if not most digital writing is intended to retain the capability of producing analogue output. This is a constraint that prevents an optimal use of the specific properties of the digital format. The benefits ascribed to 'electronic journals' are largely dependent on escaping from the analogue constraints and utilizing those properties of the digital format that cannot be achieved by analogue means.

3. Networked information

Like any technological phenomenon, digitizationis not an isolated development but one that functions within a specific technological context. More specifically, digitization involves not only the transformation of communication from analogue to digital, but also the adoption of a whole array of technical devices, systems and procedures. The technological context includes, among many others, the personal computer and the network (notably the Internet). It is the personal computer that allows information technology to function as an extension of the human body, for instance as an instrument for reading and writing. The network attaches this 'extended body' to a multitude of other individuals and groups of individuals.[13]

In order to understand the significance of the network in the context of digital scholarly communication, we might consider the body of knowledge produced in science and scholarship as represented by its accompanying set of information resources. These resources form a *conceptual* system in a number of ways due to the fact that there exist multiple conceptual relations between the resources. One type of relation is the semantic relation that can be expressed in terms of hierarchy and adjacency. A hierarchy (such as a classification – e.g. Dewey or UDC) or ontology constitutes a conceptual model of a domain onto which information resources are mapped. Another type of relation is the referential, intertextual relation between a text and the texts it cites or by which it is cited. References transform a given body of texts into a system of possibly intricately related entities, although the qualities of the relations are not visible unless they are explicitly mapped out.[14] Referential relations are similar to

[13]The concept of an 'extended body' is related to Paul Otlet's concept of the document ('le livre') as a dynamic mechanism that – in the way all machines are an extension of the human body – is also an 'extoriorization of the brain itself'. See Otlet 1934, especially p. 422–426 and Day 1997.

[14]Mapping the interrelations between texts, authors and domains by means of citation analysis is of course, one of the key methods of bibliometrics and scientometrics.

subject relations in that references are generally thought or supposed to point to resources that are of relevance to or at least related to the subject of the citing document. The difference between these two types is that referential relations are embedded in the information resources (and may be extracted to form a conceptual map), whereas subject relations are superimposed on the resources and are derived from a pre-constructed conceptual map.

In the context of analogue resources, the system is conceptual precisely because the *relations* are conceptual: they only 'exist' in the form of a map or model that is external to the information resources themselves. The system is also *static* because the resources themselves do not change and do not engage with each other through the relations. In the context of digital resources, the system of information resources differs in two respects. One is that the resources themselves need not be static: as we shall see further on, they can be dynamic in the sense that their content and external relations depend on changing contextual situations. The other is that the relations can be acted upon, e.g. one resource can 'call' or 'transfer us' to another.

It is clear, however, that for a body of digital resources to form a dynamic system they need to be embedded within an electronic network that provides the means for interaction. This explains why the use of digital documents, in scholarly communication and otherwise, is almost entirely based on networked resources rather than on 'offline' resources.[15] That is not just because the network is the more convenient means for disseminating information, but because we have come to depend on an open, dynamic system that can only exist in the context of the network. So the digitization of documents has to be seen in the context of the networked digital infrastructure in which the document is embedded, and the analysis presented in this chapter has significance only within that context.

The concept of 'networked' information implies the availability of information stored on a computer system (server) to which users have remote access over a network. In the context of scholarly information, that network will generally be the Internet, which in theory implies global availability. Global availability, in turn, implies the possibility of unlimited links between documents, i.e. between any document and any other document(s) on the Internet. Within the networked environment there need only be a single instance of an information item to allow it to be accessed globally; additional copies that are both stored and accessible over the network are, in principal, unnecessary. This

[15] Digital information need not, of course, necessarily be networked. *Offline* media such as CDROM and DVD are 'digital' in the sense that they contain information that it is encoded in binary digits, and they employ digital technology for creation and rendering. But the fusion of digital documents, personal computers and the network as an enabler of an open dynamic system is an extremely powerful force that has the potential to change the roles of the document in information and communication in fundamental ways.

in turn engages issues of ownership and control (related to both access control and control over authenticity of content) and of preservation and sustainability.

Networked information implies theoretically, as we have seen, a global presence over the network. Conceptually this means that the creator of an information resource would store it on a network node from which it can be accessed by any user, anytime, anywhere (fig. 4.1a). In the context of scholarly communication this global availability is becoming a fundamental property of digital networked information. That is to say that it is assumed (e.g. by the author and/or publisher) that a digital publication will be globally available from a controlled source on the network, and that links to any other resource on the network contained in a publication will provide access to that resource. Global accessibility, controllability of the resource and unrestricted connectivity have become fundamental characteristics that distinguish digital, networked information resources from printed, analogue resources (table 4.1).

Table 4.1. Fundamental characteristics of digital networked information resources

Globality: the resource is globally accessible over the network.

Controllability: the author or publisher holds control over content, form, functionality and availability of the resource, including the option to change or withdraw the resource.

Connectivity: any other resource on the network is a valid candidate for linking to, and can therefore be accessed through the resource.

However, even when the information chain is digital and networked, one or more of these fundamental characteristics are often constrained when network access to publications is mediated through intermediary actors in the information chain, (Fig. 4.1b). These actors may put restrictions on the extent to which information is globally accessible. For instance, publishers apply access restrictions (or require libraries to impose them) in order to protect copyrights. Libraries may store copies of publications locally, which means that authors no longer control the document as accessed by the user and that the dynamic nature of digital documents may be violated. The digital network has the potential to increase the information creators control over content, accessibility, connectivity etc. But in practice the network is not just a technically determined infrastructure but also a social system where control is negotiated among actors in the information chain. The examples given here show that there is often a tension between technical possibilities and social (including economic, legal, cultural and psychological) constraints.

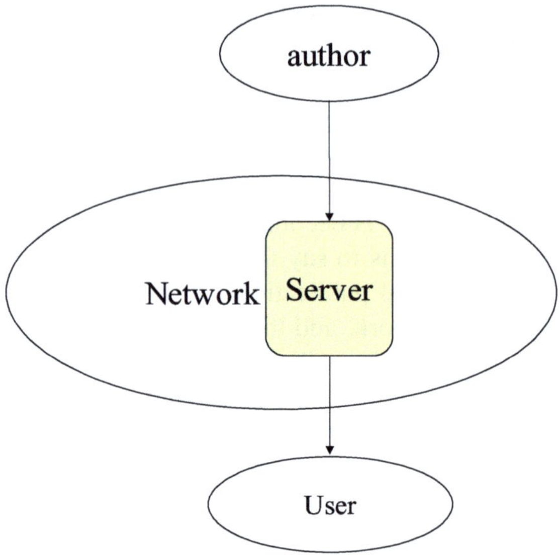

(a) Direct networking

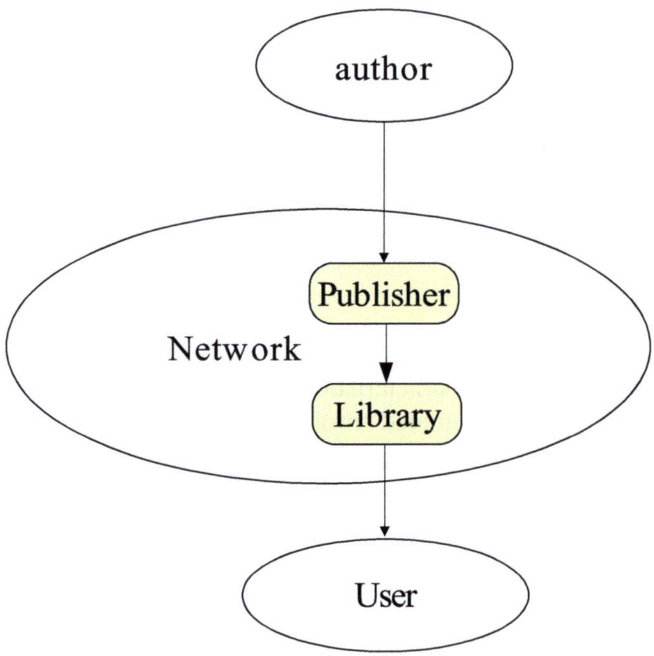

(b) Mediated networking

Figure 4.1. Global network access

4. From static to dynamic information: the mutable document

Permanence or 'fixity' is an important property of the printed medium. What has been put down on paper cannot easily be changed without leaving visible traces of the fact that the text has been altered. In the case of printed publications, there may also exist numerous copies in many locations against which an item can be checked. The fixity of the printed document is not only a safeguard against fraud. It also gives the document a certain degree of independence. Once published, printed information is not even under control of the author or publisher; it will remain an invariable statement until the final copy disappears. Only new versions, or comments on the original can attempt to modify the message conveyed by the original text. But the original text itself is fixed for once and for all – any revision or new edition can be regarded as a new document that exists separately from the original (which has no way to refer to a subsequent revision or edition). Or, in the case of materials added to the original text, as two co-existing but distinct documents consisting of the additions superimposed on the original. David Levy, in a criticism of Bolter's concept of the permanence of the printed medium versus the dynamic nature of digital texts, denies fixity by giving the example of annotating a copy of a memo as proof of the potential fluidity of printed (or written) documents.[16] But this is not a convincing example: a car with a flower pot on its rooftop is not a different car, it is in each and every respect the same car (with a flower pot on its rooftop).

The fixity of the written or printed document is in many respects a useful characteristic. It increases the trustworthiness of the document and allows it to function as evidence and as a reliable historical record. It provides a certain safeguard against fraud in a context such as scientific communication where certification plays an important role. But it can also be perceived as a limitation, exactly because it does not allow the document to reflect change. What an author has written might not reflect his or her current thinking, or it might lack additional information that is highly relevant. Or it might be the case that the data required for composing the document or even the requirements themselves are not available until the very moment that the document is to be used within a specific context.

In this respect digital documents are, potentially, fundamentally different. Because of the separation of carrier and content, changes in content are easy to apply and are not reflected in any kind of physical or technical characteristic of the document. A change in content does not affect the medium on which the information is recorded, and cannot therefore be detected without reference

[16] Levy 1994.

to an external standard (e.g. an original and certified copy, a checksum or a 'fingerprint'[17]). Moreover, digital information resources – notably resources within the scholarly information chain – are made available and accessed over the network. In principle therefore, there need be only a single source[18] that is globally accessible. Changing the single source of a resource will change every instance of that resource accessed thereafter.

This is not to say that it is an inescapable property of digital documents that they change, i.e. are *unfixed*. Following Levy (1994), we argue that a digital document is merely *mutable*, that is: it can take on any state between fixed and fluid, and this state can be changed at any time. In fact most information resources in the realm of scholarly communication are fixed, whether distributed in printed or digital form. But the point is that digital information is mutable *in principle*, and that we should therefore ask the question to what extent this property is actually used in scholarly communication.

A case in point is that digital information resources need not be designed to *record* or *document* what has been thought or found or has happened, but to *reflect* and to *monitor* a current and possibly dynamic state of affairs in real-time. Reflecting change in this way can be the very purpose of an information resource, even if it is presented in the *form* of a document. The lack of fixity is an essential property of this type of information resources and they are designed with this in mind. In some information systems documents are in fact highly contextualized, being created only at the moment of access and for the individual user requesting the resource.[19] The digital document that reflects the current state of affairs at the precise moment that it is consulted, represents a fundamental departure from the traditional non-digital document. The traditional document fixates whatever it represents in time; it carries by definition a historical representation, referring to whatever has occurred in the past. The real-time document has no historicity whatsoever, not only because it reflects the present, but also because, by implication, whatever it represents will not be preserved in that document.

Multimedia

The dynamic nature of the digital document also allows the inclusion of non-static information types such as animations, moving images (video) and sound, generally described as 'multimedia'. As such, this is often an important 'added value' property of the digital document that cannot be achieved in printed form.

[17] Morrissey 2002.

[18] E.g. the source addressed through a universal resource locator (URL) or digital object identifier (DOI). See above for restrictions due to mediated access.

[19] For instance when a document is compiled from various fragments, perhaps obtained from a variety of internal and/or external databases or other sources, and based on specifications supplied by the user. Such documents are often said to be created 'on the fly'.

But it also leads to the question in what way multimedia contribute to the infixity of the digital document. The multimedia document's content in terms of 'bits and bytes' need not change over time (i.e. the multimedia content itself is fixed). What changes is the manifestation of the content to the user. It can be argued, however, that this is true of any document, perhaps with the exception of a document containing information that can be perceived in a single glance (e.g. a written note or digital soundbite containing the message 'finish now!').

Dynamic documents

The possibility of adapting information to specific contexts and forms of use provides all kinds of opportunities for creating new categories of documents that may provide benefits in specific situations. These are cases where information is changed *intentionally, frequently* and *contextually.* This means that the information is adapted (or made to be adapted through an appropriate system or procedure) by its creator or maintainer to specific circumstances (e.g. changes in the facts reported, or different user characteristics) as often as is required in order to achieve a specific result. We use the term *dynamic* information to describe this type of information. Dynamic information cannot be described completely in terms of its content (including the form in which that content manifests itself) without a description of how that content and/or form change over time.

Table 4.2 summarizes four strategies that can be employed for creating dynamic documents. The first two (*monitoring* and *updating*) are strategies that are not applicable to printed documents and that in many cases result in new digital document types: documents that are intended to be reflections of the current state of affairs, with a time-frame varying from immediate to perhaps no more than 24 hours. Examples are continuous weather or traffic congestion reports and online newspapers.

The other two (*revising* and *expanding*) are strategies that are usually applied to existing document types, and that might well have a parallel in printed form.[20] For instance, new editions of books and revised versions of reports are common examples of these strategies. An example of these strategies applied to a a digital version of a traditional printed form is the Stanford Encyclopedia of Philosophy,[21] described by Allen et al., as a 'dynamic reference work'.[22]

In terms of the fixity/fluidity debate, the first three are especially interesting. Whereas *expansion* only adds to, but does not change the original text,

[20]Note, however, that as described above the outcome of these strategies is different in the digital world from that in the analogue world. In the former they modify the original document; in the latter they lead either to an additional, modified document, or to a second document that is only significant in combination with the original document.

[21]http://plato.stanford.edu/.

[22]Allen et al., 2002.

monitoring, updating and *revision* are dynamic updating strategies that alter the original text *in situ* and therefore contribute to the fluidity of the digital document.

Table 4.2. Updating strategies

Monitoring	Continuously adapting in real-time the content of a document to the current state of affairs. Examples: a document that reflects the current temperature; a document with a picture generated by an active web-cam.
Updating	Frequently and regularly (e.g. several times per day) adapting the content of a document to new information that has become available, e.g. in the context of an online newspaper or news site.
Revising	Occasionally changing the content of a document, e.g. to reflect changing insights of the author or to correct errors.
Expanding	Adding materials such as annotations, user comments, hyperlinks etc. to the contents of a document.

Dynamics of the user environment

The dynamic nature of digital information resources need not only be related to what the *author* wishes to define as its content and to the dynamic nature of what the resource is about. It can also relate to the dynamics of the user environment. Fixed information resources such as printed documents cannot take the differences between users into account: there is a single resource that is the same for each user. Digital documents, however, can be contextualized by adapting themselves to specific user requirements. Here too, various strategies can be followed (table 4.3). In general these strategies require some form of interaction between the user and the information system, although in some cases (e.g. with personalization) it is often sufficient to provide a one-off profile of the user to the system. The generation strategy requires input of a query by the user, possibly in the form of a query profile that is valid for a certain period. Combined with one of the strategies described in table 4.2, this would result in an alerting or 'early warning' system. The modularization strategy requires direct interaction between the user and the information system, since it is the user's choices and navigational behaviour that determine which content modules will be presented.[23] It is a strategy that has been described by Ann Peterson Bishop as 'knowledge disaggregation' in a study that describes how article components are identified, mobilized and used by students and faculty members.[24]

Table 4.3. Adaptive strategies

Personalization	Using personal characteristics of the user to define the contents of an information resource.
Generation	Using query data (as input by the user) to define the contents of an information resource (usually generated from a database).
Modularization	Using the navigational behaviour of the user to select components of an information resource.

Dynamic documents in the context of scientific communication

In the previous chapter we discussed an approach in the context of e-science, where the research process might be monitored in the information chain during the throughput-stage. This would imply that scientific documents dynamically monitor the state of research. We have argued, however, that the requirements for certification make this a less likely scenario, at least as far as formal scientific communication is concerned.

Nevertheless, the formal scientific article does become potentially more mutable in the digital environment. There are three forms in which mutations may emerge (table 4.4). Firstly, it is relatively easy to link a document to new versions that effectively supersede the original, even though the original remains in place. A second form is the addition of new materials (annotations, comments, references etc.) that expand the original document. Finally, the text of the document can be revised by means of changes *in situ*.

Another form of mutation happens when digital documents cited in an article change, possibly without the knowledge of the citing author. To the extent that such citations perform argumentative functions, the citing author can no longer be certain of the fixity of his or her argumentation. To put this differently: the mutation of documents, especially by means of expansion and/or revision, has not only implications for the document that is changed, but potentially also for any document in the citation network in which it is included. The consequences of modifications to any document in a highly networked,

[23] See also Section 5 107 ff.

[24] Bishop 1998.

Table 4.4. Mutation forms

Versions	Documents that are superseded by a new version or "edition" with its own identification (e.g. version or edition number) without deleting or changing the original or previous version
Expansions	Documents that retain their original contents and structure, but are expanded, e.g. through the addition of annotations, commentaries, appendices, additional hyperlinks etc.
Revisions	Documents of which the contents are replaced in whole or in part by new content, erasing the original content

and hyperlinked context are therefore potentially far more significant than in the analogue world. They are also uncontrollable due to the fact that a given document has no knowledge of other documents that provide links to it.

The dynamic nature of digital information also provides a number of opportunities and possible drawbacks for the various actors in the information chain:

- The author can benefit from the dynamic nature of digital information in various ways. One is by creating new 'real-time' document types that continuously reflect a state of affairs, rather than by creating a sequence of documents that only reflect discrete points in time. Another option is to use a document as a 'crystallization point' (or even as a portal) to which revisions, annotations, user comments, hyperlinks etc. can be added. Or an author might wish to keep a text on a certain topic abreast with his or her current ideas about that topic and to eliminate ideas with which he or she no longer agrees. In an extreme case, the author may even withdraw a text by removing the only official copy from a server, effectively making it cease to exist. All these things are possible, but do require active 'management' by the author. However, it also requires a degree of control over the copy or copies that are available to users. As already mentioned, this may be problematic – for instance if there are intermediary actors such as rightsholders (e.g. publishers) that do not cooperate, or libraries that store copies locally.

- For the publisher, dynamic online documents are a means to maintain relations with customers who have to return to benefit from changes in documents. This offers many opportunities for value-added services, but it also requires setting up provisions for maintenance. An added benefit for the publisher as a rightsholder is that the more dynamic a document is, the less incentive there will be to infringe property rights through copying, since a copy would become outdated and lose its value.

- The user might be able to obtain correct and/or timely information from the document if it is indeed kept up-to-date. However, it might also be difficult for the user to ascertain the level of fixity of a document, and he may therefore be unsure as to its timeliness and authenticity. For instance, the user might not be able to ascertain whether it is the original version as initially published, or a later, updated version. As a consequence, it might also be difficult to ascertain whether a document that is accessed through a resource locator (e.g. a hyperlink) is identical to the one supposed to be referenced by that locator in another document.

5. Quasi-intelligent documents

In the previous section we have looked at a specific and important property of digital documents: their potential mutability or *infixity.* Infixity, however, remains a property that depends on an external operation, on a change that is *applied to* the document. In this section we shall look at a number of properties that are *intrinsic* to the digital document.

Digital documents are capable of performing what appears to an observer to be some form of 'intelligent' behaviour. They are capable of adapting themselves to the user, carrying out specific tasks at the request of the user, and entering into relations with the outside world. This behaviour can be described as *quasi-intelligent*, following the argument by Turing (1950) that behaviour can be regarded as intelligent if a human observer cannot decide whether the behaviour is human or machine-based.As Harnad 1992 argues, Turing Test intelligence is a goal that cannot as yet be met. Nevertheless, digital documents do appear to show a certain form of behaviour that may be regarded as 'intelligent' by the user, and therefore we use the concept of intelligence here in the metaphorical sense of 'quasi-intelligence'.

Human intelligence is characterized by our ability to acquire knowledge both about ourselves and about our environment, allowing us to reflect on ourselves and the outside world, to make decisions, adapt our behaviour to prevailing circumstances and to take appropriate action. Documents that have been constructed with a view to intelligent use demonstrate a similar set of properties. The concept of 'intelligent' Dovey (1999):

> 'The application of the object-oriented or document-centric paradigms to current metadata practices leads to the concept of intelligent documents, where the document not only contains the metadata tags detailing the construction and content of the document, but also the application code for interpreting and manipulating this metadata. In effect the document is introspective in that it understands itself. The use of virtual machine concepts would enable such application code embedded in the document to be platform independent, and agent-oriented paradigms would enable the document to take autonomous action.'

Dovey also gives a number of examples of applications of such intelligent documents:

- Platform and application independent metadata 'tagging', allowing the document to 'understand' itself and its own structure and content.
- Documents that can autonomously assist the user in navigating, understanding and even manipulating its own content.
- Documents that can autonomously communicate with other intelligent documents, for example in automatically organizing themselves when placed in an object-oriented database or container, or in dynamically establishing and maintaining links with other relevant documents.

The reflexive document

We usually regard documents as linear texts that we peruse in order to obtain information about a subject and to follow the argumentation of the author. For instance, in reading a newspaper, we can read the text in the linear order in which it has been written and printed. There are, however, other options open to the reader, such as to skip sections or to read them in a different order. Many documents – most non-fiction books for instance – are constructed in such a way as to offer further functionalities, such as a table of contents for navigating through the text, and an index to locate specific topics.

Including a title page, table of contents, index, internal references and bibliographical references within a document makes it reflexive: it contains knowledge about itself and its semantic and intellectual context, and it is capable of operating on itself. This allows us to ask a document questions about authorship, the location of a specific topic or the source of a particular idea. The document provides the answer by supplying an author's name or an internal of external reference. Documents are, of course, a very special type of physical object in that they contain knowledge of a world exterior to themselves. What we argue here is that they also contain knowledge about themselves that in the digital realm allows them to *utilize* this 'external knowledge'.

This primitive form of 'intelligence' is not unique to digital documents – it is also a characteristic of traditional print documents. But the possibilities are greatly enhanced through digitization. Embedding documents in software (or software in documents) allows them to be used as programmes that perform tasks in reaction to external stimuli provided by the user (e.g. a mouse-click). This form of intelligence (which in practical terms manifests itself as *functionality* – see section 6) is an important condition for the dynamic and interactive nature of digital documents. The most obvious example of this is of course *hypertext* that allows a document to respond to a stimulus (provided by the user) by instructing a software system (e.g. a browser) how and where to find other information it refers to. Another example of a document that contains knowledge of itself that allows it to take specific action is e-mail. Replying to an e-mail message requires no more than a simple mouse-click. There is no need for the user to find or even know the e-mail address of the recipient, or even his or her identity. The document itself contains all information required, and is coded in such a way that it can pass this information on to a software system that will carry out the requested transaction.

Standards for document encoding such as SGML, HTML and XML are examples of how reflexivity is embedded within documents. They define explicit, codified semantics that provide the document with knowledge about its structure and the role of textual components. Such documents contain not only their primary informational content, but also reflexive information about the document itself. In practical terms this reflexivity can be used for including active (and perhaps automatically derived) navigational devices within the document (e.g. linked tables of contents and indexes, thumbnail-overviews, navigational histories, etc.).

The role of metadata

Such reflexive information is usually described as *meta-information* or *metadata*. Traditionally, metadata has always been *added* to the informational content of documents as a conscious act by an author, editor or cataloguer. In this sense, documents are not intelligent by themselves, but are made intelligent. In the analogue world, metadata about the informational content is 'internalized' within the document itself (e.g. front matter, contents tables, indexes etc.). Metadata about the document itself is also 'externalized' e.g. in bibliographic descriptions contained in bibliographies, inventories and catalogue entries. Systems for managing and retrieving analogue documents are generally based on the latter type of metadata; in fact these are systems that are based on metadata representations and therefore operate almost by definition in absence of the physical documents they describe.

It is interesting to note that if we define bibliographies and catalogue entries themselves as documents that contain knowledge about other documents (i.e. the documents they describe), and if we accept that they are also reflexive in the sense described here, they become agents that are capable of performing certain actions with respect to those other documents. We have defined bibliographic descriptions as 'external metadata'. If these are now internalized within the documents they describe, we obtain documents that are self-sufficient in that they can function without any need for external metadata. Given an appropriate software environment, documents containing 'embedded' metadata can function intelligently (e.g. by responding to a request for information on a specific topic or by a certain author) without further support. The fact that digital documents no longer necessarily require external intervention such as cataloguing is of some significance. The creation of external metadata and related systems (such as manual and computerized catalogues), necessary in order to manage and retrieve documents within a controlled environment such as a library or archive, has historically been a significant function of these 'memory institutions'. Digital documents can form self-organizing systems that can take over the metadata functions traditionally performed by these institutions.

Semantic linking

Following a hyperlink or answering an e-mail are rather primitiveexamples of the 'intelligence' of digital documents. With ordinary hypertext, the document knows no more than that there exists some relation between itself and another document. It has no knowledge of the link's semantics, and cannot inform the user about the *nature* of the relation. A solution for this problem can be found in the application of 'typed' links: hyperlinks that include information about the nature of the relation. In a study on modularization of documents, Harmsze has developed a link typology that provides explicit labeling of all internal and external references (fig. 4.2).[25] Through these typed links, the document can inform the user about its internal logical structure as well as about its external relations and/or adapt its behaviour, e.g. the form of presentation or choice of modules. This allows the document (and the reader or software that processes the document) to distinguish between links to resources that – for instance – support or deny a certain argument.

The principle of typed links has also been included in the XML-standard and especially within X-Link.[26] This may eventually lead to a much richer and more expressive use of hypertext and eventually to the establishment of a 'semantic web'.[27]

Adaptive hypermedia

Another example of documents that can adapt themselves to the user is what are generally called *adaptive hypermedia*.[28] Adaptive media are designed to create a model of the user and what he or she already knows about the subject covered in the document. They will then adapt themselves to provide the knowledge required in that specific context, i.e. they are capable of *self-contextualization*. Knowledge about the user (in the form of a formal

[25]Harmsze 2000. In the context of modularization the use of typed links is essential in order to be able to navigate between the various components or modules of a document. See also Kircz (1998; 2001; 2002).

[26]Cf. W3C-Consortium 2001.

[27]See for example (Berners-Lee et al., 2001; W3C, website; Sumner and Shum, 1998) and the 'Research in Semantic Scholarly Publishing' project of the Erasmus University Rotterdam (http://rssp.org). An example of the use of semantic linking to enhance the argumentative function of scholarly communication is given in Uren et al., 2003.

[28]Brusilovsky 1996; Brusilovsky et al., 1998; Bra et al., 1999.

1. Organizational relations
   - Hierarchical (Is part of, Contains)
   - Proximity (Article, Project, External)
   - Range (To narrower range, To wider range)
   - Administrative (To meta-information, To contents)
   - Sequential (Back, Next)
     - Representational (e.g. Text, Figure, Animation)
2. Scientific discourse relations
   - Communicative relations
     - Elucidation (Is elucidated in, Elucidates)
       - Clarification (Is Clarified in, Clarifies)
         - Definition (IS defined in, Defines)
         - Specification (Specification given in, Gives specifications)
       - Explanation (Is explained in, Explains)
     - Argumentation (Is argued in, Provides argumentation for)
   - Content relations
     - Dependency in problem solving (Depends on, Is used for)
       - Transfer (Input from, Output to)
     - Elaboration (Is elaborated in, Elaborates)
       - Resolutions (Is detailed in, Is summarized in)
       - Context (More context in, Is focused in)
     - Similarity (Agrees with, Disagrees with)
     - Synthesis
       - Generalization (Is generalized in, Is specialized in)
       - Aggregation (Is aggregated in, Is segregated in)
     - Causality (To cause, To effect)

Figure 4.2. Typed links (from Harmsze 2000, Appendix A)

representation of the user's world view) is acquired either by direct input or by deduction from the user's behaviour. The document is capable of recording the user's interaction, i.e. his or her reading and navigational behaviour. It can also reason about that behaviour to infer which components are required to serve the user's information needs. Taken together with the technique of modularization described above, this might lead to *automatic modularized contextual adaptation.* In this case the semantic information about the link-type is matched against the user model to decide the trajectory through the document. The process can be totally transparent and requires no active user-involvement: the document appears to respond intelligently to the context in which it is used.

Embedded software

Modularization, semantically typed links and adaptive hypermedia are examples of methods for adding intelligence to documents in order to align content, structure and presentation to the requirements of the user. This resembles the way an intelligent person will not give everybody the same information about a certain topic, but will adapt the information to the level and background of the listener. Documents that adapt to the context by aligning themselves to the user can be said to possess a certain degree of social intelligence.

But intelligent documents can do more. The examples given so far depend for their intelligent behaviour on a software environment, e.g. browsers, e-mail clients and hypermedia systems. But there are also examples of documents that contain software within themselves.[29] The most basic examples are Word documents with embedded macro's, and web documents with embedded JAVA scripts. These documents contain not only the data, but also the software required to perform a certain task.

In scholarly information, embedded software can be used to include computer models, simulations and visualizations. This allows, for instance, a user to compute a formula and visualize the results in different ways while reading a scientific paper. Or a report on survey research could link to the underlying data sets and allow the reader to perform additional analyses on the data. This is especially important in view of the increasing role of databases in scientific research. A well-known example is the depositing of research data in very large genomic and proteomic databases[30] that is required for acceptance of publications in the field of molecular biology.[31] Embedded software could provide a transparent link between research publications and the underlying datasets in these cases.

A further example of the use of embedded software in documents has been provided by Robert Wilensky and others from the University of Berkeley.[32] They introduce the concept of 'multivalent documents' containing embedded

[29]It might be argued that what we describe here as 'intelligence' in documents differs from human intelligence in that it always requires *some* form of software and hardware environment to operate. Even documents with embedded software functionality require a computer with an operating system to run on. But it can also be argued that human intelligence cannot function (or does not even exist) without the physiological context of the human body, the social context in which it operates, and – in our present times – a sophisticated technical infrastructure.

[30]Such as the GenBank (U.S. National Center for Biotechnology Information) and the Protein Data Bank (European Bioinformatics / Research Collaboratory for Structural Bioinformatics).

[31]Brown 2003.

[32]Phelps 1998; Wilensky 2000. See also the Multivalent Browser website at http://http.cs.berkeley.edu/~phelps/Multivalent/.

intelligence that provides functionality in a way that is totally transparent to the user. In other words: the document 'reacts' in certain ways without the user having to call a specific function. A (theoretical) example would be the case where a reader is consulting a document that contains an image of a 17th century French publication. By selecting an area of the text, it is automatically replaced by a text in English and in a modern typeface. The underlying processes (including optical character recognition and automated translation) are completely transparent to the user. The intelligence of the document is defined here by its ability to express itself in various languages.

Another example of the intricate relation between content and software is the Stanford Encyclopedia of Philosophy mentioned earlier. Allen et al. (2002) describe this type of document as a D[ynamic] R[ererence] W[ork] that 'includes a highly customized workflow system by which the members of an entire discipline are empowered to collaboratively write *and maintain* a refereed resource. Such a resource would not only introduce traditional topics in the discipline but would also track the (new) ideas that are constantly being published on those topics in a variety of media'.

Documents as interface

When a digital document is both dynamic and networked, it is no longer merely a container of information, but can also be regarded as an access point for, or *interface* to any type of global networked resources. For instance, a digital report might contain a diagram of a chemical process. Clicking on the diagram might open a window that shows a simulation of the process, running on a supercomputer thousands of miles away, based on parameters contained in the document. The user might then click on a reference to a database in an entirely different location, which would send new parameters to the supercomputer and so modify the simulation. The user could then modify the report based on the new findings and send it to a colleague.

In this way a document operates as an intelligent interface to a variety of networked systems and resources. Because of its knowledge of the outside world, it can dynamically link resources that might not even have any knowledge about each other. The example given above is in fact but an extended example of a common type of resource found on the Internet: the 'home page' or 'portal'. These are resources that contain little or no content in the traditional sense, but that solely function as an interchange between the user and available resources. Although similar to printed guides or bibliographies, the digital version is different in that it actively switches the user to another resource, or – to put it in more pragmatic terms – retrieves a resource for the user.

Control mechanisms

So far we have described the intelligence of digital documents in implicitly positive terms as an added value property that would benefit the user by meeting his or her specific requirements, offering more functionality or even empowering the user above the author. However, the impact of digitization is not always beneficial in the sense that the term 'added value' suggests. This is due to the fact that technology is never an agent of change in its own right, but always operates within a social context. People will use and abuse technology, in ways intended or not intended, expected or not expected by its creators. Intelligent documents are often, as we have seen, software products or at least containers of embedded software. Since it is a characteristic of this type of document that it can acquire knowledge about the outside world, including knowledge about the user, it is also capable of harming the user. One example is its capability of recording user behaviour, information preferences etc. and of passing that information on to third parties, perhaps without awareness and/or consent on the side of the user.

Another issue is that an intelligent document is capable of making decisions. It can therefore also decide what actions are allowed, by whom, in which context and when or for how long. This allows a high degree of control by the author or other rightsholder, especially if the document is encapsulated within a digital rights management environment (see section 8). The use of such protective control mechanisms is, of course, related to issues concerning the economic interests of rightsholders and is facilitated by recent legal developments that allow rightsholders greater control as well as protection of the technological protective measures used.[33]

It is interesting to note that in this way a kind of 'reversal' is taking place between the public and private domains. Public information, i.e. information created and/or held by public institutions has always – despite the use of the term 'public' – been notoriously difficult to access for those outside the institution. Although this still is the case in many parts of the world and with many individual institutions, there are also many examples of an increased 'openness' of public institutions reflected for instance in legal arrangements that provide individuals and organizations with certain rights to obtain information from government bodies. This openness is also visible in the increasing amount of information made available by public bodies through the Internet, and in many forms of 'e-government' that create a larger level of transparency than before. On the other hand, publishers, as private organizations, have traditionally been champions of 'openness' providing access to the world's knowledge through their publications. Openness is, one might ague, the very substance

[33]Sonneland 2001; Strickland 2003a,b.

of publishing, and publishers have always embraced new developments that were seen to provide opportunities for better access to and larger distribution of content (related, of course, to larger turnover and profits). It is therefore significant that digitization leads to what may be called 'the closing of the document' – to documents with embedded control mechanisms that intentionally limit the extent to which they can be accessed: by whom, in what way, in which environment and within which time-frame.

Documents as a dynamic system

As we have seen in the previous chapter, the traditional body of printed scholarly publications can be represented as an essentially *static* and conceptual system. Printed documents are fixed nodes that are related to other nodes, explicitly through citations and/or conceptually through a superimposed descriptive system such as a classification or ontology. The body of interrelated scholarly publications is, in the phrase of Fytton Rowland (1997), the 'canonical archive' of science.

The body of networked *digital* documents, on the other hand, can be regarded as an open *dynamic* system. Dynamic systems are characterized by the fact that the components and/or their relations change over time. Open systems are characterized by the fact that they can be influenced by events outside of the boundaries of the system. A further distinction can be made between *adaptable* systems (systems that are changed by an external actor, e.g. through a graphical user interface) and *self-adaptive* systems (systems that modify themselves in response to perceived changes in the environment, e.g. user-input or changes in the internal composition of the system). The system of networked digital resources is both adaptable and self-adaptive. It is adaptable in the sense that it can respond to external inputs (e.g. an author updating a resource). It is self-adaptive in the sense referred to above in our discussion of quasi-intelligent documents.

Table 4.5. Dynamic information resource systems: outputs

Static content	Rendering of stored content with no (self)adaptation
Dynamic content	Rendering of content as a result of (self-)adaptation
Links	Transfer to other resource nodes within the system

An open system requires inputs from and outputs to its external environment (Tables 4.5 and 4.6). In the case of scientific information resources, the outputs are notably the renderings of resources (i.e. digital documents) within a user interface, usually as a response to some user-input requesting a set of resources, or one or more links to other resources (i.e. a transfer to another node in the system). Renderings may be either static (i.e. based on fixed content) or dynamic (i.e. based on adaptation by the user or self-adaptation by the resource).

In the context of dynamic networked resources, the inputs are rather more complicated because we have to distinguish between different types of actors. The inputs provided by authors or publishers (or more generally: those holding maintenance rights) involve revising the content, form or functionality of an existing resource. The user is involved in interactions with the system that lead to the outputs described above. And finally, changes in the internal or external environment can act as an input to specific nodes of the system. For instance, a resource might reflect additions to the system by incorporating links to these resources, or it might reflect changes in the external environment which the resource can monitor through a sensing device. (e.g. a resource reflecting the current state of a laboratory experiment).

Table 4.6. Dynamic information resource systems: inputs

Revision	by author or publisher
Interaction	by user
Reflection	of changes in the internal or external environment

6. The functional document

The concept of '(quasi-)intelligence' as related to information resources is elusive. It describes a characteristic as perceived by a human observer and seems to denote the fact that digital resources *appear* to be (or to act) like intelligent beings, or at least more so than non-digital resources. It does not refer to the nature of the digital resource as such, but rather to our perception of it. What is then the intrinsic nature of this 'intelligence', what is it that makes the digital resource seem to show intelligent behaviour? The key lies in the word *behaviour*: the digital resource is able to engage, communicate, interact, adapt itself, in short: to 'behave' or to *act*. Michael Buckland has described this characteristic of the digital document in terms of *functionality:*

> 'Fifty years ago, one would look up logarithmic values in a printed book of 'log tables' in order to do calculations. The volume of log tables was a conventional document. Today, one ... would use an algorithm to compute log values as needed. The answer given should be the same. The table and the algorithm seem functionally equivalent. What has happened to the notion of a 'document'? One answer is that whatever is displayed on the screen or printed out is a document. One might say that the algorithm is functioning as a document, as a dynamic kind of document. ...

> It would be consistent with the trend ... towards defining a document in terms of function rather than physical format'.[34]

In this example the functionality of the digital document has taken the place of discursive content: the functional 'document' does no more than provide access to and display the output of a function. In terms of systems analysis, this functionality is equivalent to a procedural change of the documents (knowledge-)state based on internal rules and user-supplied parameters.

The functionality of the log computing document is but one example of what seems to be a more fundamental characteristic of the digital document. Other examples – some of which we have already discussed – include:

- documents where content is created on-the-fly from database records;
- adaptive documents where content is based on the document's knowledge of the user context;
- documents that provide data or audiovisual sensor outputs in real-time[35];
- documents that connect the reader to other resources (the content of which may not be known when the connecting document was created).

The content and form of a digital document therefore may depend on (i.e. be created or modified by) its functionality. What distinguishes these examples from the log computing example given by Buckland is that they are highly contextualized, e.g. with regard to time and user. The log value computing document operates on a static relation between input (a number) and output (its log value). The contextualised examples given above are based on dynamic relations with multiple parameters that result in outcomes that are a priori unpredictable.

Drawing on our analysis in the preceding sections, we can describe the digital document as an information resource that is potentially *dynamic* (in that its form, structure and content can change in a number of different ways), that is capable of acting as a node in an open networked system, and that is capable of acting on the basis of internal knowledge and external stimuli. The concept of a *functional* document can now be summarized as follows (table 4.7). First, functional documents are *interactive:* in order to do anything useful, they will have to be able to communicate with their environment and to interact with users and systems. Second, the functional document is *reflexive,* it has knowledge about itself that it can use either to adapt itself or to interact with the

[34]Buckland 1998, p. 215.

[35]Consider for instance the transformation of a traditional genre such as the travel guide. Rather than providing static data, narrative and pictures, the digital document can provide not only moving images and sound, but also real time data (temperature, web-cam images) and contextualized data (a list of hotels tailored to the user's profile).

environment. Third, functional documents provide a certain degree of connectivity, that is they act as an interface for the user to other resources. Finally, functional documents are capable of adaptation, i.e. of changing their internal structure and /or contents depending on context, internal knowledge and external stimuli.

Table 4.7. The characteristics of functional documents

Interactivity	the ability of a document to engage in interaction with the environment, notably users and software systems.
Reflexivity	the ability of a document to interact and/or adapt itself based on its own internal characteristics.
Connectivity	the ability of a document to act as an interface to other digital resources on the network, or to accept a connection from another resource.
Adaptability	the ability of a document to adapt its structure and contents to specific contextual characteristics and requirements, notably those of the user.

Since functionality is a direct outcome of the digital nature of the document, the extent to which this characteristic is utilized defines what may be described as the document's *digitality*.[36] A static document that is no more than a binary encoded copy of an analogue document can therefore be said to be less 'digital' than one that shows the characteristics described in table 4.7.

7. The 'copy paradox'

Another quality of the digital document is the extent to which it can be copied, i.e. its *replicability* (Levy, 1994). This is related to two further characteristics: the separation between carrier and information (in the generic sense of a construct embodying content, form and functionality) and the representation by means of a binary encoding. The latter allows a bitwise perfect copy to be made of the information, that is: a copy that is identical in every respect to the original. The former means that the carrier is of no consequence for the documents content and functionality, but merely a matter of convenience or convention.

Now it is clear that it is essential for a copy that it can be displaced, removed from the location of the original. Combining the two qualities of networkability and copyability, we now have the situation that it is extremely easy to create

[36] Note that the term *digitality* has been used in a different sense by other others, notable as an equivalent of binary opposition or *binarity* by Baudrillard (1983, p.115, 145).

a copy of a digital resource and to transfer it to a different location. Ultimately, since 'a bit is a bit', any *rendering* of a resource is in fact the creation of a copy (i.e. a bitwise representation of the original). What is then essential is the ability to store the bits used for rendering. But again, since 'a bit is a bit', there is in theory no difficulty involved in doing so. In the digital world, in order to use information, one has to render it. And rendering it means that there is a digital copy. Since the original can be stored, and the copy is identical to the original, the copy can be stored as well. The only difference is the location at which it is stored.

This is also, of course, a problem. Since it is the very nature of the use of digital information that the user obtains a copy of the original, the user is empowered to store and/or re-distribute that copy as if it were the original. Protection of copyrights has therefore become extremely important, but also extremely difficult, requiring sophisticated technical systems and complicated legal measures that put severe restraints on the use of digital resources (Strickland, 2003a).

At first sight it would seem that in a networked environment there is no need for storing and redistributing copies of information resources, as long as access to these resources is open to all users without any restrictions, financial or otherwise. But given the fact that in most cases the costs of creating and maintaining information resources have to be recovered, 'free access for all' is not a realistic proposition. There are at least three reasons why rightsholders go to significant lengths to prevent 'unauthorized' copying by imposing restrictions on access. One reason, and in most cases the most important one, is the fear of redistribution of the copy. The recent cases of peer-to-peer systems that have led to large-scale redistribution of copyright materials are an example of the consequences of the properties of digital information resources that we have described here.

The other reason is that the rightsholder may wish to prevent re-use of the information by the original user. This is an issue when the rightsholder changes the business model from the traditional 'purchase' or even 'licensing' model to a 'rental' model (e.g. based on 'pay-per-use'). Within such a model, income depends on the frequency of use, which means that charging for repeated use by the same user is profitable.

Finally, the author may wish to retain control over content and functionality of the document, e.g. in order to incorporate modifications in a dynamic way that reflect new facts, changing opinions etc. This becomes impossible when the user substitutes a local copy for the original source.

In all these cases it is paramount to prevent the user from making and storing a copy, to prevent re-distribution, re-use or substitution by the user. As a consequence, traditional 'legal exceptions', i.e. exemptions from legal limitations on copying for specific user groups or types of use (e.g. personal study), tend to become eroded.

A further question is why users wish to make or use copies. In simple terms, if information resources were always easily available at no cost over the network from their original source, there would be little incentive to making and redistributing copies. However, since there are costs involved, and access might not be guaranteed, there is not only an incentive to make copies, but also to redistribute them as a favour to or in exchange for other resources with other users. An additional incentive to seek and use copies might be a lack of access rights to the copied resources.

What we conclude from this, is that the problem of copyright violation a a complicated issue. Digitization has empowered the user to obtain control over the digital resource. Rightsholders are attempting to regain control through technical and legal measures. However, whereas for rightsholders copyright is predominantly a financial issue, this is not entirely so as far as the user is concerned. Other issues involved include access, convenience and social considerations. There is a paradox here in that if information were to be available over the network free of access restrictions and free of cost (as advocated by movements such as Open Archiving), copying would not be a problem, but would not be necessary either.

There is, however, another issue that is related to the fundamental nature of digital resources as discussed in this chapter and that has not often been considered. Although the problems of extensive copying and copyright violation seem to be a consequence of digitization, it can also be argued that they are a result of the fact that the properties of the digital resource are not fully utilized. As we have seen, these properties include the embedding of resources in an open dynamic system, based on the concept of functional resources that interact and change. Ultimately, such resources cannot function in isolation from their operating environment: an isolated, static copy would be or soon become inferior to the original, if not totally useless. If digital resources are designed in such a way that they lose their interesting properties outside of their operating environment, the problem of unauthorized copying might become irrelevant.

We argue therefore that the extent of copying is not only determined by the factors described above, but also by the degree in which the properties of the digital networked resource are utilized. The more dynamic and functional a resource is, the less incentive there will be for copying.

8. The problem of authenticity

We have defined dynamic information as information that changes *intentionally* in form and content over time. It is here that we find one of the important differences between analogue and digital information: digital information can be changed without any difficulty. This can be done with good intentions or with bad intentions, in ways that highlight modifications or disguise them. It can be done by whoever is responsible for the content, and it can be done by anyone else. It can be a function of the control over the medium exercised by the author, or a function of a control strategy allowing the reader to construct the information content. This problematizes the concept of authenticity and the means for actors in the information chain to guarantee that a document is the document it claims to be, and that it has not (been) changed in form and content since its creation. It is almost impossible to ascertain the authenticity of digital information without special and complicated technical measures such as encryption, signatures (such as checksums), digital watermarks[37] and steganography,[38] securely archived reference copies, and secure document formats such as PDF.

The concept of authenticity in the context of dynamic documents is in fact rather more complicated. Digital documents need not necessarily 'exist' prior to their consultation by a user. They are often created at the very moment of consultation, e.g. when the generative strategy is used whereby form and content depend on various parameters including some provided by the user. This means that even at a given point in time T_n there may be no unique 'authentic' document, but a multitude of versions each of which is only authentic in the context of a single user at a specific moment. In this sense the dynamic document is often merely a *potentiality*, a system of possibilities that materialize depending on time and context. Which of these possibilities will actually materialize cannot be predicted, and the result, the sum-total of all materializations, will never be seen or recorded.

It has even been argued that sometimes the dynamics of digital resources have more significance than the contents. An example of this is the online newspaper as described by Sheila Thiel (1998):

> 'More and more, the popularity of an online newspaper is based on its 'fun' quotient, its interactivity, the ability it gives readers to 'click through' from one link to the next. There is no need to absorb content in an online newspaper, just to revel in the gestalt of easy travel through the words. The online newspaper is not a product, it is an experience.'

However exaggerated this view may be, and however unrealistic in the context of scientific communication, it does reflect the extent to which the dynamic digital document can in fact be constituted in the reading experience of the user rather than through the writing of the author. The act of reading, or – more precisely – of navigating through a (possibly dynamic) information resource, need not lead to communication of the author's message to the user, although it might well lead to the creation of content in the user's mind. The fluidity of the digital format and the potential dynamics of the user environment described above reinforce the constructivist nature of communication.

[37] Petitcolas and Kim 2003.
[38] Katzenbeisser and Petitcolas 1999.

It is especially the contextualization and personalization of dynamic information resources that problematizes authenticity, precisely because it transfers control from the author or rightsholder to the user. Ultimately, when information is a product of time, context and user, the concepts of authenticity and authorship cease to exist.

Authenticity in scientific communication

For the archivist the concept of 'authenticity' is important because most archival documents have legal implications (e.g. offering proof of a transaction) and are also more or less unique: only the original (with perhaps a limited number of certified copies) is regarded as the authentic document. Authentication and certification are expressed through specific tokens such as signatures, seals and stamps. In scientific communication, however, certification is derived from the context, i.e. publication in a recognized scientific journal. The 'authentic' document is not the author's manuscript submitted to the journal, but any copy of the definitive published version.[39]

Here, the concept of authenticity is not related to 'originality', but to the fact that the document (e.g. the research article) is 'contained' within the journal. The 'evidence' or 'authority' provided by a document as used in scientific discourse is not based on 'the document at hand' as is the case with archival documents: the scientific author can *refer* to another document without having physically to produce that document to support his or her case. This is possible as long as the document referred to belongs to the 'canonical archive' of science and can be assumed to be universally available for scrutiny by fellow scientists wherever they may be, and at any point in time. This assumption is based on the fact that the scientific communication system – using printed publications as its main distribution medium – guarantees that scientific documents such as research articles and monographs are produced in relatively large quantities, that all copies are equal, that they are distributed throughout the academic world, and that they are preserved and held accessible by libraries.[40]

[39] At least this is the view held by publishers, as witnessed by a recent press release from *Nature Publishing Group* (10 January 2005):

> 'This policy [with respect to self-archiving by authors - JMO] ... is also designed to protect the integrity and authenticity of the scientific record, with the published version clearly identified as the definitive version of the article.' (http://www.macmillan.com/10Jan2005NPG.asp)

It remains to be seen whether the growing importance of self-archiving and repositories will ultimately challenge this concept of authenticity.

[40] The production and distribution of multiple copies of texts has been an important factor for the development of science since at least the early Middle Ages, and is the reason why a large majority of ancient texts have survived (Cisne 2005).

These three factors – containment, referral and universality – are characteristic of formally published documents and of the scientific communication system based on such documents. Together they explain the fact that the *archival* concept of 'authenticity' is not seen as an important issue in scientific communication and in the world of publishing and libraries.[41]

The possible consequences of digitization of the scientific communication system (e.g. modularization, dynamic documents, single source rather than multiple copies) does, however, lead to concerns about authenticity.[42] Until now, these concerns are largely linked to the issue of digital preservation, where the concern is mainly about preservation strategies that modify the original publication (e.g. by migrating to new formats) and therefore challenge the authenticity of the document.[43] These modifications are generally necessary to cope with ongoing developments in hardware, software and standards.

Safeguarding authenticity

The consequence of the potentially dynamic nature of the digital document is that one of the key functions of analogue documents, viz. their role as *evidence*, as proof of a state of affairs or a certain conduct, cannot be taken for granted in the digital realm. This is of course problematic in a domain such as scientific communication where certification plays an important role. Specific measures with respect to the processes of creation and maintenance of digital information are required, as well as technical constraints on formats, media and access mechanisms. These issues are currently addressed both in the domain of digital archiving and preservation,[44] and in the area of copyright protection.

In digital archiving and protection the key problem is, as we have seen, to preserve the authenticity of the document over time when technological developments and economic contingencies demand some form of transformation of the medium that acts as a container for information content. Emulation is an approach that seeks to avoid this problem, but it is uncertain that this will provide a viable solution in the long run.[45] A guarantee that 'bitwise authentic' copies can be rendered after perhaps centuries is difficult to imagine. An

[41] Although a librarian will, of course, check the authenticity of a valuable or unique book that is offered for sale in the same way that an archivist would do. Of course, a publication may reflect inauthenticity in the *research process*, such as misrepresentation of data, plagiarism etc. See LaFollette 1992.

[42] See for instance Greene and Cockerill 1997. A Committee on the Dissemination of Scientific Information (CDSI) workshop held in 1998 recommended to the International Council for Science (ICSU) 'that ICSU approves and recommends a system for ensuring and maintaining the integrity and authenticity of electronic publications as well as devising means for promoting uniform standards.' (Shaw and Elliott 1998)

[43] The issue was first brought up by two seminal reports, one by the Task Force on Archiving of Digital Information (1996) in the USA and another by Mackenzie Owen and van der Walle (1996) in Europe.

[44] See also footnote 72 on page 75.

[45] See Rothenberg 1999 and Bearman 1999. For a more balanced approach to migration see Mellor et al., 2002.

additional problem is due to the fact that the rendering of a document (i.e. the representation that is seen by a user) is not only determined by the document, but also by the software used for rendering. This is already the case with e.g. html, where the exact rendering is determined by the underlying 'html engine' of the browser.

Various large-scale projects have been set up to develop solutions for the preservation of digital materials such as the CEDARS project,[46] CAM*i*LEON,[47] NEDLIB[48] and DNEP.[49] One of the outcomes of these projects is that full preservation of all aspects of a digital document is illusionary, and that one has to base preservation strategies on so-called 'significant properties' that are both essential and preservable.[50]

Current developments in technology applications that might serve to protect documents against fraudulent use (including both intentional misrepresentation and copyright violations) are mainly found in the area of copyright protection with a focus on digital rights management[51] and copyright management systems.[52] DRM systems take two approaches to securing content. The first is 'containment', an approach where the content is encrypted in a shell so that it can only be accessed by authorized users. The second is 'marking', the practice of placing a watermark, flag, or a XML tag on content as a signal to a device that the medium is copy protected.

The Electronic Privacy Information Centre describes Digital Rights Management systems as follows:

> 'Digital Rights Management (DRM) systems restrict the use of digital files in order to protect the interests of copyright holders. DRM technologies can control file access (number of views, length of views), altering, sharing, copying, printing, and saving. These technologies may be contained within the operating system, program software, or in the actual hardware of a device.'[53]

In other words: the issue of using technology for maintaining authenticityis mainly regarded in the context of control strategies that are developed to safeguard economic rather than intellectual interests. This use of technology, however, still presents many problems, including a lack of trustworthy computing devices, robust trust management engines, and a general-purpose rights expression/authorization language,[54] and there is increasing end user concern

[46]http://www.leeds.ac.uk/cedars/.
[47]http://www.si.umich.edu/CAMILEON/.
[48]http://www.kb.nl/coop/nedlib/.
[49]http://www.kb.nl/hrd/dd/dd_onderzoek/dnep_ltp_study-en.html and http://www-5.ibm.com/nl/dias/preservation2.html.
[50]CEDARS Project 2002a,b; Yeung 2004. See also Diessen and Werf-Davelaar 2002.
[51]Gervais 1999; Martin et al., 2002.
[52]Becker et al., 2003; Picot 2003.
[53]Electronic Privacy Information Center (2004).
[54]LaMacchia 2002.

about restrictions on their access to, and use of, information.[55] It has also been argued that all digital protection schemes can be broken, and that therefore neither copyright nor authenticity can be protected in the domain of digital information.[56]

9. Reading, creating and control

So far we have considered the concept of a dynamic document in terms of a single document that might take on different identities. There may be a finite number of discrete identities (e.g. different versions or editions) or a potentially infinite number of context dependent identities (e.g. when documents are created 'on-the-fly' or reflect a continuously changing state of affairs). The modularization proposed by Harmsze and Kircz[57] is an example where a document is deconstructed into semantically typed components, from which the user compiles his or her own version, depending on interests, available background knowledge, etc.

But the adaptive strategies described in table 4.3 offer a level of control that goes beyond the boundaries of the individual document. Since in the digital world document boundaries are no longer fixed, and information is structured as a multitude of interlinked fragments, the user becomes the creator of his or her individual reading experience. He or she has a choice in navigating through a web of links and in setting the boundaries between what is to be considered a document and what belongs to another document. What is a 'document' is in fact defined by what the user perceives as a coherent set of interrelated items of information. Increased application of internal and external hyperlinks, modularization and the use of semantically typed links could increase the creative behaviour of the user in constructing a personal and unique reading experience. Although most authors will attempt to create a well-defined and coherent document, many readers will end up by creating their own document from (fragments of) various author's documents. It is in fact not at all clear whether there is (or at least need be) any question at all of authorship of the content, meaning and message perceived by the user in such a dynamic and creative reading experience. It is more likely that such content fragments are perceived in terms of external facts and (objective) 'data' rather than as subjective expressions by authors with individual human identities.

Therefore the reader of digital documents is in a different position – and in a different relation to the author – than the reader of printed documents. With the printed document the reader is confined to what the author has in store. It is the fixity of the printed medium that empowers the author to control the

[55] Foroughi et al., 2002.
[56] Schneier 2001.
[57] Harmsze et al., 1999; Kircz 1998; Kircz and Harmsze 2000; see also page 110.

information process. In the digital world the reader or 'user' can become the author or 'creator' – he or she is in control of form and content, and ultimately of message and meaning. It is interesting that earlier observers of the Internet as a technical medium do not seem to have recognized this. For instance, Thompson (1990, p. 164-171) describes *fixity* (together with *reproduction* and *participation*) as one of the key attributes of technical media (without, of course, referring explicitly to the Internet). Fixity, defined as a medium's capacity to store information, is related to structures of power and control. The argument is, in short, that the vast storage potential of technical media empowers those that use that storage capacity, offering them control over what information is available to whom. Shapin illustrates this with the example of the Ford Motor Company's website:

> 'The website allows the organization to store vast amounts of corporate information about itself, its products and services ... The webpages allow the company to control precisely which information is stored and made available ... It provides the company with a unique forum for generating a display of trustworthiness and integrity'.[58]

Although Shapin does recognize potentially negative effects (including user dissatisfaction with an uneven balance of power), he seems to neglect the way the Internet empowers the *user* (or, in Shapin's example, the consumer) to ignore whatever are the intentions of the author (or producer) and create his or her own 'store of information' through a personal choice of content and form. It is precisely the *infixity* of digital information that empowers the user. The only way for the author to regain a certain amount of control is to decline to use the distinguishing properties of the digital format such as hyperlinks and other navigational devices. But by doing so, the digital format would lose its significance and the author would become less attractive for the user who has come to expect these properties in the digital environment.

In sociological terms public information spaces (such as the scholarly information chain, public broadcasting or the Internet) can be characterized as power spaces in which control is disputed between producers and consumers or authors and readers. There are multiple dimensions to this dispute, including the rights/profits dimension and the dimension of control over and impact of the messages conveyed through the information space.

On a metalevel there is also a struggle *about* the power structure of information spaces. This is a struggle that always seems to follow a distinct pattern. At the introduction of a new technical medium, arguments in favour of the medium are offset by the argument that it empowers the *sender* (e.g. the author, tv-producer, creator of computer games, or in more institutional terms 'the government' or 'industry') to an extent that endangers society in general

[58] Shapin 1996, p. 63–64.

and freedom in particular. The idea is that the user will not be able to escape the force created by the 'power' of the medium (a power that is, of course, also ascribed to the sender using the medium). Ultimately it is usually found that the power balance, if it was at all disturbed by the medium, soon becomes restored. This is due to the fact that the user is always less disempowered by the medium than expected. Users are not informed by media, they use media to construct information. In general, new and technical media empower the user because they expand the opportunities for this type of construction, e.g. by expanding the amount and accessibility of alternative sources. This has been seen with television (e.g. with the proliferation of tv channels) and it is significantly the case with digital networked media.

We argue therefore that one of the key consequence of digitization (and of the infixity of the digital medium) is a shift from reading of fixed documents created by an author to reading of documents (in terms of content, form and functionality) that the reader himself has created. In contrast with what has often been argued then, the digital format does not empower only the author or producer of information, but also the reader or consumer. In general terms, whereas the author or publisher determines the conditions under which information resources are made available, it is the user who has the power to use (or perhaps abuse) those resources to construct his or her individual reading experience (table 4.8).

Table 4.8. The power structure of the networked information space

Author / publisher	the power to determine which information resources are available to whom, and under which conditions.
Reader / consumer	the power to determine content and form of what is actually read.

The author of a digital document is therefore in a precarious position. He or she can use the digital format merely as a carrier for traditional forms and genres, as a surrogate for the printed form. The author then maintains full control over form and content of the publication, and authorship and authenticity are assured. But many specific properties of the digital format will not be utilized, and the resulting degree of innovation is low. On the other hand, the author can make full use of the intrinsic properties of the digital format. The degree of innovation is high. But in that case, far more control is given to the reader, and authorship and authenticity may lose their significance.

10. Characterizing the digital scientific article

10.1 The digital document

The characteristics of digital information that we have described in this chapter allow the document to be much more than just a 'computer-readable' version of the traditional, analogue document. As we have seen, digital documents can contain explicit knowledge about themselves and their users (e.g. through metadata, tags, typed links etc.). They have the potential to function as open entities, embedded within a network of other documents of which they have knowledge and to which they can refer. They may contain in-built intelligence due to the fact that they contain or even consist of software that allows them to perform tasks that are relevant to the user. As a consequence, digital documents can acquire knowledge about the user and adapt content, presentation and functionality accordingly – just as the user can also interactively adapt the document. Therefore, a digital document need not be the same for each and every user.

Another important conclusion to be drawn from our analysis is that digital documents can be regarded as *software* products. This implies a theoretical distinction between analogue and digital documents due to the fact that digital documents cannot be described merely in terms of content and form, but need to be described in terms of behaviour or *functionality* as well. We have also seen that the digital document is characterized by its dynamic nature, i.e. by changes in content, form and behaviour over time. A further implication is that digital documents require, as all software, continuous management and maintenance.

Our characterization of digital documents as (potentially) *intelligent* has an important consequence for authors and the process of creating documents. Documents do not become intelligent by themselves: it is the author who has to insert intelligence into the document, even if there are tools to assist in this task. Authors therefore would have to have some knowledge of the technical issues involved, they would have to acquire the skills required to create what is essentially a software product, and doing so would have to become embedded in the culture of scientific writing. Also, as documents become more closely interlinked and interdependent, authors would also have to consider the networked environment in which they operate. And if content and functionality of a document depend on the quality and availability of other documents and information resources, new responsibilities for and relations between authors become necessary. Finally, digital scientific documents, once created, may require continuous maintenance by the author.

For the reader too, digitization potentially creates a different situation. Digital documents require a more active involvement of the user, as can already be seen with the use of hyperlinks where the user has to make decisions and adopt a conscious navigational strategy. The more dynamic, interactive and intelligent documents become, the more readers will have to accept the insufficiency of a linear reading behaviour. On the other hand, readers could benefit from the enhanced functionality and the extent to which documents can adapt themselves to his or her characteristics and requirements.

Finally, both authors and readers will have to engage in the legal consequences of digitization. Issues concerning licenses and access rights are a direct consequence of the embedding of software functionality in the document and of interfacing to networked resources. A document is of no use to the user if he or she has no rights to address its functionality and/or linked resources: it simply will not work, and will cease to act as a carrier of information. As we have seen, it is in this area that control becomes an important aspect of the digital document. The dynamic, interactive and intelligent possibilities of the digital document are technically and – within a scholarly context – economically feasible. The extent to which mechanisms can be found to strike an acceptable balance between the various aspects of access and control is certainly one of the factors that will determine the future development and acceptance of these possibilities.

10.2 The digital scientific article

In this chapter we have analyzed the potential characteristics of the digital document in fairly general terms, with little reference to specific document genres. Applying this analysis to the digital scientific article, we might expect to find at least some of the following characteristics:

Multimedia content: the article contains, in addition to text, various forms of multimedia content such as colour images, moving images, and sound.

Network access: the article is globally accessible at a networked storage location.

Network connectivity: the article contains active links to a variety of resources such as cited literature, background information and data resources.

Author control: specific characteristics such as single-source storage and direct access (i.e. or the lack of intermediated access through publishers, libraries etc.), but especially mechanisms for revising the published text allow the author to exert control over content and availability of the article after its initial publication. Editorial and copyright policies of the journal are also relevant in so far as they allow or restrict author control.

Dynamic content: the article employs a variety of strategies to allow dynamic content, e.g. to monitor changes, adapt to new information, reflect changing insights and to include additional materials.

Adaptability: the article employs a variety of strategies to adapt its form, content and/or functionality to the context in which it is used, including characteristics of the user.

Functionality: the article shows advanced forms of quasi-intelligent behaviour by means of such devices as navigation mechanisms, semantic linking, adaptive hypermedia, embedded software, interfacing etc.

Copyability: the article employs a conscious strategy for copyright protection either implicitly through the use of dynamic content and functionality, or explicitly through the use of digital rights management devices.

Reader control: the article employs a variety of the above-mentioned strategies to allow the reader to create his or her individual reading experience.

Flexibility: the article is not constrained by journal-related factors such as periodical publication dates, article-length, lay-out, amount of graphical elements, etc.; also, one would expect greater flexibility – than with printed journals – in terms of submission formats, forms of peer review and copyright policies.[59]

Potentially, therefore, the consequences of digitization can be significant, not only in general but also for the scientific article. In the next chapter we shall investigate the extent to which these characteristics are actually to be found in articles published in electronic journals.

[59]This characteristic is not the result of digitization of the scientific article as such, but rather a consequence of digitization of the publication process that frees the article to a certain extent from its meta-structure, the scientific journal.

Chapter 5

THE ELECTRONIC JOURNAL 1987-2004

In this chapter we move from theory to reality by looking at the actual impact of digitization on the scientific journal. We begin by developing an analytical framework based on the concepts of digitization described in the previous chapter. This is then used as the basis for an empirical study of peer-reviewed e-journals from the first stage of development ranging from 1987 to 2004.[1] A second study examines the use of digital properties in open access journals from the third development stage. The chapter ends with a summary of the results and conclusions.

1. Analytical framework

The analysis of digitization presented in the previous chapter allows us to develop a framework for determining the extent to which digitization has impacted on the article as a medium for scholarly communication. The research model on which this study is based, is described in fig. 5.1.

Our approach is based on a two-level analysis:

1. The aggregate level of the journal and journal collections (e.g. the fonds of a publisher or the offerings of an electronic content provider). At this level we look at the technical properties that are made available or allowed by the editors or publisher, as well as at the editorial policies that govern characteristics such as style, typography, structure etc. This level provides the context in which the individual article is embedded, and represents the

[1] The development stages are described in section 5.

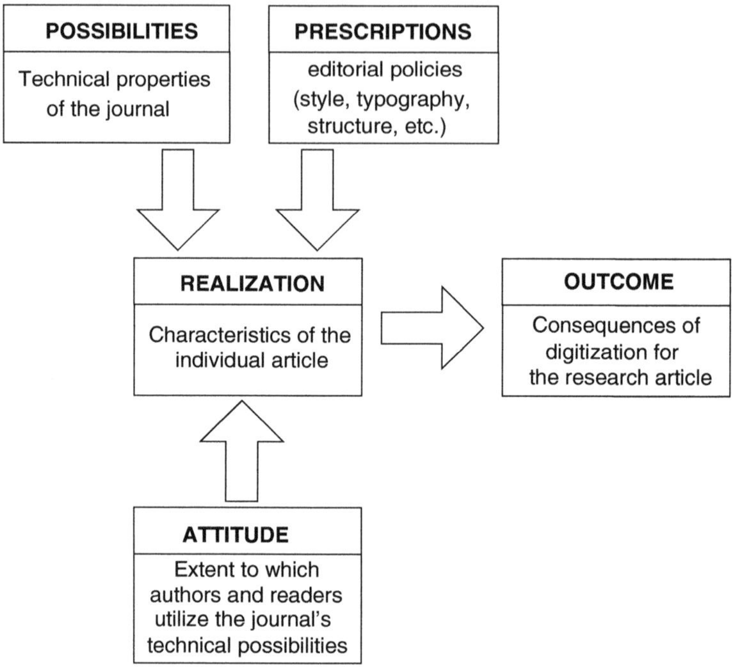

Figure 5.1. Research model

viewpoint of the editors or publisher providing a publication platform as a communicative link between author and reader. For instance, a journal may allow and encourage inclusion of multimedia, but may also prohibit their use.

2. The level of the article itself, viewed as the product of an author or group of authors. At this level we look at properties as used by authors in their articles, i.e. the extent to which the journal's possibilities and ambitions are realized. For instance, authors may decline to use multimedia, even if they are allowed and encouraged by the journal.

The product of what the journal allows or prescribes on the one hand, and the attitude of the author towards utilizing the journal's 'digital' properties on the other hand, is visualized in the characteristics of the individual article. By examining a larger number of e-journals and their articles, we have sought to obtain an overall view of the consequences of digitization for the scientific article. In order to describe the characteristics of articles in e-journals, we have identified a number of technical properties that are listed in table 5.1. We have examined the presence of these properties in the journal, and the use that is made of them by authors in their articles.

Table 5.1. Article properties

Formats	Includes the format(s) in which articles can be submitted by the author, and the format(s) in which articles are published by the journal. Formats include HTML, word processor formats, (La)Tex, PDF, etc.
Multimedia	A property that is related to the question of format, is the extent to which multimedia information (e.g. images (photographs, drawings), video and sound) is used for conveying scholarly content. We would expect to find a wide range of embedded and linked multimedia objects to enhance the expressive power of the digital article to convey scholarly information.
Data resources	Provisions for embedded or linked datasets etc. related to the publication.
Revision	Mechanisms for revision by author, updates, forward references etc.
Response	Mechanisms for addition of user comments, etc.
Customization	Mechanisms that adapt the representation of content based on user characteristics. At the aggregate level: functions such as 'My Journal' where users can maintain profiles that govern the selection, format etc. of articles, or store selected articles in a personal file system.
External hyperlinks	The inclusion of external hyperlinks (url's) and whether they are active (clickable); the addition or activation of hyperlinks in the editorial process; also special cases (e.g. typed links).
Functionality	Any type of quasi-intelligent mechanisms such as semantic linking, adaptive hypermedia, embedded software, interfacing, simulations, automatic translation, etc.
Navigation	Special means for navigating through the text at the article level, or through the journal or journal collection (aggregate level), e.g. linked content tables, navigation bars, topic maps, etc.
Peer review	Approaches to peer review that exploit the interactive properties of the digital format.
Copyright	Copyright policies that take into account aspects of the digital format.

Table 5.2. Editorial policies

Style rules	Enforced style rules such as APA, MLA etc. and the use of proscribed article structures.
Limits	The extent to which the journal imposes restrictions on article length, amount of figures, etc.
Typography	The extent to which the journal prescribes typographical characteristics, e.g. spacing, fonts, headings, margins, page size.
Digitality	The extent to which editors describe and encourage the use of digital properties (rather than presenting the digital journal as a new distribution mode for a format that in other respects is similar to the printed format).

As mentioned above, the extent to which journal articles reflect the characteristics of the digital format is not only constrained by these technical properties, but also by the explicit editorial policies that apply to digital journals. For instance, the required use of traditional style guides, typographical conventions, conventional article structures etc. limits the freedom for the author to use new modes of expression and visualization that are made possible by digital formats. Limitations imposed by strict editorial policies can force the journal to resemble a traditional print journal. On the other hand, editorial policies can also acknowledge the novel characteristics of the digital format, and editors can encourage authors to use these characteristics in order to distinguish the digital journal from its printed form. Therefore, we include in our analysis a number of elements related to editorial policies (table 5.2).

In summary, therefore, we have analyzed e-journals and their articles both with respect to the possibilities and prescriptions controlled by the journal and its editors/publisher, and with respect to the actual practice as performed by authors in their articles, in order to arrive at more general conclusions with respect to the digitization of the scientific article.

The relation between the article properties and the characteristics of digitality are summarized in table 5.3. This table reflects the way we shall move from our descriptive analysis of e-journals to the more general conclusions with respect to the impact of digitization, based on the characteristics of the digital scientific journal listed at the end of the previous chapter.

Table 5.3. Research design

Digital characteristics	**Article properties**
Multimedia content	Multimedia
Network access	(All E-journals studied are globally accessible over the network).
Network connectivity	Formats, hyperlinks, data resources.
Author control	Revision, editorial policies, copyrights.
Dynamic content	Multimedia, revision, functionality.
Adaptability	Formats, functionality, navigation, response, customization.
Functionality	Functionality.
Copyability	Formats, copyright.
Reader control	Formats, customization.
Flexibility	Submission formats, peer review, copyright, editorial policies.

2. Research data

For this study we have used a sample of e-journals that was created on the basis of an extensive search using the sources listed in table 5.4, Internet search engines and various references from the literature.

Journals were selected using the criteria listed in table 5.5.

The objective of these criteria was to select those e-journals that can be expected to maximize the use of the digital properties of the e-only journal rather than journals that use the Internet merely as a distribution channel or to develop new business models. We have therefore excluded journals that are just digital versions of existing print journals,[2] and also journals published under the Open Access model (e.g. by BioMed) because we believe this to be a separate category where the use of the digital format is primarily focused on the new business model rather than on innovation of the journal itself.[3]

A total of 186 e-journals were selected based on these criteria. Theseare listed in the appendix at the end of this chapter. It is difficult to establish the

[2] Note that print journals that are available in digitized form through projects such as JSTOR (http://www.jstor.org/) and MUSE (http://muse.jhu.edu/) were therefore not included.

[3] This assumption is checked by means of a separate analysis of open access journals in section 4.

Table 5.4. Sources of e-journals

Title		URL
ARL Directory of scholarly electronic journals and academic discussion lists	–	http://db.arl.org/dsej/
Cyberlinks - Innovative e-journals	–	http://www-97.oeaw.ac.at/cgi-usr/ita/cyber.pl?cmd=get\&cat=29
Directories of Electronic Journals Beyond UCSD	–	http://gort.ucsd.edu/ejourn/jdir.html
Directory of open access journals	–	http://www.doaj.org
EJI(sm): A Registry of Innovative E-Journal Features, Functionalities, and Content	–	http://www.public.iastate.edu/~CYBERSTACKS/EJI.htm
E-Journals.org	–	http://www.e-journals.org/
CIRS - International Centre for Scientific Research	–	http://www.cirs-tm.org/
Internet Public Library	–	http://www.ipl.org/div/serials/
Mathematics on the Web	–	http://e-math.ams.org/mathweb/mi-journals.html
NewJour	–	http://gort.ucsd.edu/newjour/
Penn Library	–	http://www.library.upenn.edu/cgi-bin/res/sr.cgi?resourcetype=17
WebJour - Scholarly Journals Distributed Via the World Wide Web	–	http://info.lib.uh.edu/wj/webjour.html

exact amount of existing journals that would meet our criteria, and to ascertain how comprehensive this sample is. In view of our extensive search, however, we believe that our coverage is fairly complete and certainly representative. Llewellyn et al., (2002), in a study of the use of e-only journals in scientific research, found 144 journals that satisfied their criteria for inclusion.[4] A survey by Hitchcock et al., (1996) identified only 35 e-only journals in the area of science, technology and medicine. In a study of the viability of early e-journals, Crawford (2002), using slightly different criteria (e.g. only free journals, but including journals with print version), found a total of 104 refereed journals

[4]These criteria were similar to those used in our study: currently being published, peer reviewed, original publication in electronic format, research oriented, 2-3 years of archive (min. 20 articles) available, contain English language articles, and no subscription required.

Table 5.5. Selection criteria

Peer review	Only journals that included a statement about peer review were included in the sample in order to focus on the issue of formal scientific communication. This excluded many informal newsletters, creative journals etc.
E-only	Only journals that do not have an equivalent printed version were included.
Start date before 2002	The set of e-journals was limited to those that had their first issue before 2002. This was done to allow a journal a certain time to become settled and for authors to adjust to the possibility of incorporating digital properties into their articles.
Still in operation	Journals were excluded that have ceased publication, as far as that could be ascertained. The reason for this was that apparently unsuccessful journals might not be illustrative of normal practice and might therefore distort the results of our analysis.

being published in 1995, of which only 49 were still operative in 2002.Since our database includes the same number of journals published in 1995 or earlier (and still operative in 2004), we conclude that our set is at least as comprehensive as that used by Crawford.

It should be noted that many of these journals have secured a place in academic practice. Llewellyn et al., found that two thirds of the e-only journals identified in their study were being indexed by major indexing services, scientists were finding these new publications and citing them, and libraries were providing access to them through their online catalogs.

Data was gathered for each journal based on editorial information (e.g. journal mission statements, guidelines for authors, etc.) and at least 10 articles published by the journal. In general, the analysis is based on the most recent volume. Older volumes were studied when the most recent volume contained less than 10 articles. More articles were studied when required to obtain sufficient data on the various characteristics. For instance, if the most recent issue contained many articles with a large number of multimedia, the journal was scored as a journal containing multimedia. However, if few multimedia were

Table 5.6. All disciplines

Discipline	Number of journals	% of discipline
Humanities	41	22.05
Social sciences	40	21.50
Law	7	3.76
Sciences	86	46.24
Other	12	6.45
Total	186	100

found, and it was the editorial policy to include multimedia, a larger number of articles were checked.

3. Results of the survey

3.1 Scientific fields and publication year

Table 5.6 shows the overall distribution of scientific fields in the sample, tables 5.7-5.9 show a further breakdown per discipline. The sciences account for almost half of the journals found. In fact, it would be more precise to say *only* half: of the ca. 8.500 journals covered by the ISI indexing services, 66% are in the sciences, 14% in the humanities and 20% in the social sciences.[5] In contrast with what is often thought, the humanities and social sciences have contributed significantly to the early development of the digital journal.[6]

The breakdown per year is given in figure 5.2. Figure 5.3 gives a further breakdown per discipline. These figures show that there was a peak in the start-up of e-journals during the second half of the nineties, with a sharp increase from 1995 onward, and an almost equally sharp decline after the turn of the century.

[5] As of January 2005, with thanks to Henk Voorbij.
[6] See also Nentwich 2003, p. 109 ff.

Table 5.7. Humanities

Discipline	Number of journals	% of discipline
History	6	15
Musicology	6	15
Language & Literature	15	36
Other	14	34
Total	41	100

Table 5.8. Social sciences

Discipline	Number of journals	% of discipline
Sociology	9	22.5
Psychology	5	12.5
Political science	5	12.5
Economics	5	12.5
Education	11	27.5
Communication	4	10
Other	1	2.5
Total	40	100

Table 5.9. Sciences

Discipline	Number of journals	% of discipline
Medicine	18	21
Biology	7	8
Physics	12	14
Chemistry	11	13
Mathematics	24	28
Technology	8	9
Other	6	7
Total	86	100

The type of journal in terms of publisher is often not at all clear for many e-journals. On the basis of our sample we estimate that about 40% are run by one or more individuals, 25% sponsored by academic institutions, less than 20% by scholarly societies, and only a very small percentage by commercial publishers. It should be noted that changes are not infrequent, e.g. journals

1st vol.	Number	Percentage
1987	1	0.54
1990	2	1.08
1991	4	2.15
1993	9	4.84
1994	9	4.84
1995	24	12.90
1996	34	18.28
1997	29	15.59
1998	29	15.59
1999	25	13.44
2000	13	6.99
2001	7	3.76
Total	186	100

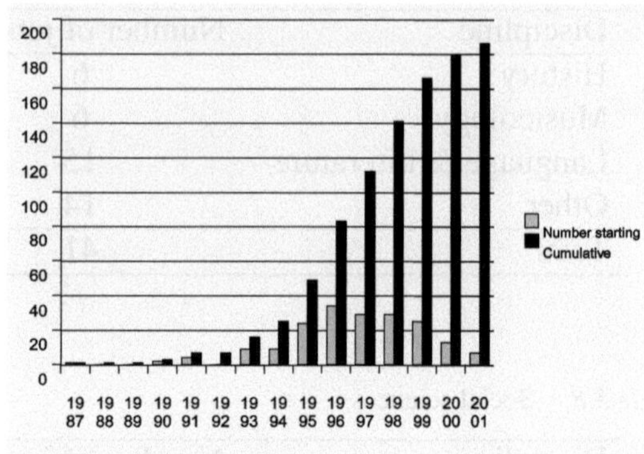

Figure 5.2. E-journals per year

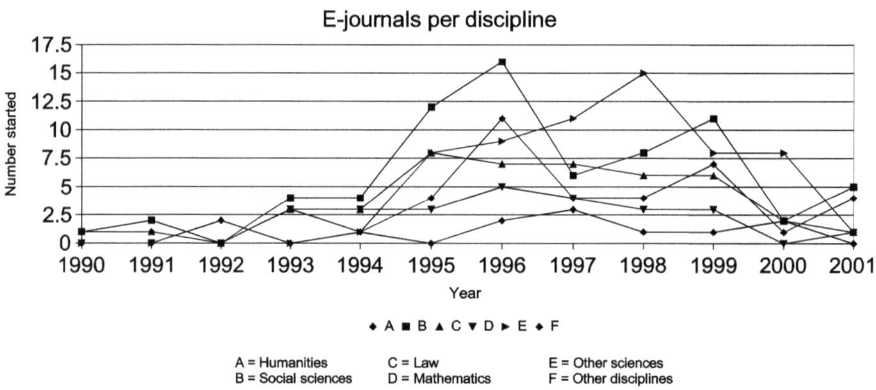

Figure 5.3. E-journals per discipline

moving from the individual to the institutional level, or being taken over by commercial publishers. On the whole, though, the early e-only journal is a small-scale enterprise set up and run by individual academics to enhance scientific communication within their own particular domain.

3.2 Submission formats

The prescribed or allowed submission format is a key variable in our study, since it reflects how the e-journal expects authors to cope with the digital format. The results are given in table 5.10. Plain text is accepted in a limited number of cases; in these cases it is clear that that no digital enhancements are expected from the author. Over 60% of the journals accept papers in word-processor formats, usually Microsoft Word and occasionally Word Perfect or Rtf. Here too, it would seem that authors are addressed in a traditional way that does not encourage a shift towards advanced digital formats. This also applies to the requirement for submitting in a (La)Tex-based typesetting format (usually in combination with specific requirements for styles and macros), although this can be used to generate digital output formats such as html and pdf, or even audio.[7] Latex and pdf are used almost exclusively in mathematics and the sciences.

In some cases the author is required to submit the manuscript in the format in which it is to be published, e.g. html or pdf. The (relatively infrequent) requirement to submit in pdf is more or less equivalent to the requirement for camera-ready copy, since the pdf format does not, in principle, allow for further editing.

In most cases, however, journals allow some degree of freedom in submission formats; submissions are then usually edited and converted to the journal's publishing format by editorial staff. This editorial process often emulates that of print publishing, leaving little room for the author to exercise control over the final output:

> 'Authors are requested to keep the text as plain as possible if Word or WordPerfect format is used; i.e., use a minimum of layout, avoid styles, etc., in order to facilitate the conversion process.' (EJCL)[8]

> 'The main text should be sent in as a non-formatted Word file (single spaced, flush left). In other words, you should not use a hyphenation program, a footnote program or any special printing formats.' (FQS)

> 'In general, manuscripts should not be submitted in html (hypertext markup language) format.' (WCR)

On the whole, e-journals do not expect authors to use novel techniques for writing and submitting manuscripts. By far the greater majority either request a relatively 'flat' input that is transformed to the journal's publishing format through a more or less traditional editorial process, or a 'camera-ready' input that is either published as-is or transformed into html. In the latter case, authors are usually required to use strict formatting rules, often dictated by prescribed macro's or styles.This will become even more apparent later on when we discuss editorial guidelines (section 3.14).

[7] An option mentioned explicitly by the Journal of functional and logic programming. See also Raman 1994 and http://www.cs.cornell.edu/Info/People/raman/aster/demo.html

[8] Codes such as 'APS' refer to the journal codes that identify the e-journals listed in Appendix A.

Table 5.10. Submission formats

Submission format	Number of journals	% of all journals
html + others	41	22
html exclusively	6	3
word	110	59
wp	42	23
(la)tex	36	19
pdf	10	5
text	25	13

Table 5.11. Publication formats

Publication format	Number of journals	% of all journals
html	125	67
html only	81	44 (65% of all html)
pdf	92	50
pdf and html	42	23
pdf only	50	27 (54% of all pdf)
postscript	20	11
dvi	8	4
(la)tex	5	3

3.3 Publication formats

The extent to which the journal and its editors are instrumental in transforming the author's submission into a form more suitable for the digital environment can be seen by comparing table 5.10 with the overview of publication formats in table 5.11. Here we see that the most frequent publication formats are html and pdf. In most cases article submissions have been converted to these formats through the publication process. However, this does not necessarily mean that in every case the author has provided 'traditional' input, of which the publication is a simple digital copy. Sometimes the the author's submission is enhanced (e.g. by activating textual hyperlinks[9]) and even including multimedia components (e.g. audio files) provided by the author.

[9]By 'activating' we mean the transformation of a network address from mere text to an item that will operate as a hyperlink when clicked on by the user; see also section 3.9.

Table 5.12. Types of multimedia

Multimedia type	Number of journals	% of all journals
Video	21	11
Audio	9	5
Software	7	4
Animations	5	3
Other	16	9

Perhaps the most conspicuous finding is the frequent use of pdf as a distribution format, either as the main journal format or as an alternative for the html publication. This is especially interesting because pdf, in spite of its hypermedium and multimedia properties, is predominantly a print-document based format.[10] Further inspection of our data reveals that 2/3 of the use of pdf as a publishing format can be ascribed to journals in the sciences. Of the 25 mathematics journals in our sample, 22 are not published in html but invariably in pdf, often with parallel versions in postscript, dvi or tex. In physics, too, the majority of journals is published in pdf only. Pdf is also often used in biology and medicine, but there usually in combination with html. The frequent use of pdf in the sciences, and especially in mathematics and physics, reflects the long-standing practice in these domains of using the (La)Tex system for scientific writing. (La)Tex is normally converted to postscript which increasingly is being replaced by pdf.

3.4 Multimedia

Figure 5.12 gives an overview of the various types of multimedia found in e-journals.[11] The percentage of e-journals offering multimedia in the various disciplines is listed in table 5.13. Although the percentages are relatively low in any discipline, both the sciences (including mathematics) and the humanities seem to have adopted the use of multimedia to a certain extent. In the legal and social sciences the use of multimedia is virtually absent. This is probably easy to explain from the nature of the disciplines. Law and to a lesser extent

[10]Nielsen 1997a, 2001, 1997b; Peek and Pomerantz 1998.

[11]We have excluded normal B/W or colour photographs, which were found in 29 (16%) of the journals. Photographs are more frequent in the humanities (27%) than in the sciences (14%) and social science (12.5%). Colour (e.g. for photographs or charts) was used in 47 (25%) journals. There are examples of a more prolific use of (colour) images (e.g. diagrams and photographs) where their use would be prohibited by cost in a print publication. See for instance the use of hundreds of colour images for a medical photoessay in the Dermatology Online Journal (DERMOJ, see Huntley 1995).

Table 5.13. Multimedia per discipline (% of journals with multimedia)

Discipline	Number of journals	% of discipline
Humanities	11	27
Social sciences	5	12.5
Law	0	0
Sciences	23	27
Other	3	25
All	42	22.5

the social sciences are apparently more text-based than other domains. The sciences can benefit from simulations, animations, stereo photographs, 3-D visualization etc. that are made possible by the digital format. In the humanities we find film and music files as the predominant multimedia types, in addition to more novel features such as virtual reality models.[12]

The use of multimedia is limited, especially taking into account that the frequency of multimedia *within* a journal is often very low, and in a number of cases no examples were found even where the journal encourages the use of multimedia (see also chapter 5.27 on page 179). For example, the Journal of Interactive Media in Education (JIME) contains only a single example of multimedia use in both volume 1999 and 2000. Of the sixteen articles published in volume 9 (2003) of the Journal of Seventeenth-Century Music (JSCM) only three contain audio attachments. Although Palaeontologica Electronica (PE) encourages manuscripts that employ animations, 2D and 3D modeling techniques, online access to databases, and online data analysis tools, very few examples were found in its articles.

Most journals that do accept multimedia enhancements allow various types, often aiming at a 'media-rich experience':

> 'Our mission is to disseminate scientific information worldwide, taking full advantage of the electronic publication medium by offering 3D graphics, video, interactive figures, ancillary databases, and sound. Special Feature Articles contain highly interactive features or large databases.' (APS)
>
> '... authors are encouraged to take advantage of the expressive possibilities afforded by JCMC's multimodal, web-based format. Articles may contain any combination of text, tables, graphics, animation, or audio components. Innovative forms of expressing research, and/or linking members of the scientific community, are welcome.' (JCMC)
>
> 'Submissions should take advantage of the multimedia capabilities of the World Wide Web, that is use audio, graphics, or video and preferably integrate text and multimedia. Text-only manuscripts will be considered inappropriate.' (MA)

[12]See Billing 2004 in *Early modern literary studies for a virtual reality model of Inigo Jones's Barber Surgeon's Anatomy Hall.* The model is available online at http://www.shu.ac.uk/emls/si-13/billing/pictures/barber2.wrl.

> 'Unlike traditional print-based journals, Palaeontologia Electronica is highly graphical in both format and content. Authors are encouraged to make use of color in their figures and tables, and to include high-resolution digital images as illustrations. Moreover, Palaeontologia Electronica encourages manuscripts that employ animations, 2D and 3D modeling techniques, online access to databases, and online data analysis tools.' (PE)

A number of journals contain extensive editorials that outline the characteristics and possibilities of the electronic format, sometimes there is also an 'example article' to illustrate what can be achieved.[13] However, even when they do accept multimedia, some journals are reluctant to leave behind traditional publishing models:

> 'Authors are encouraged to provide auxiliary material at any time in whatever formats are in current use. We shall use our discretion as to the nature and length of these auxiliary publications. This material will allow some of the unique advantages of electronic formats to be realized and also to allow the journal to adapt quickly to new developments without affecting the main publication, which is intended to exactly mirror conventional publication.' (GEOTOP)

In summary, we found the use of multimedia in e-journals to be very limited, and far more so that one might expect from the encouraging statements by some journal editors. Although many journals do provide for multimedia features, the vast majority of authors do not feel inclined to move to a more multimedia-based mode of writing.

3.5 Data resources

Examples of data resources included in or linked to by e-journals are listed in table 5.14. Only 17 (9%) journals in our sample allow for inclusion of data resources. But even in these cases, articles containing data resources are relatively rare. For instance, the LMS Journal of Computation and Mathematics (LMS) encourages authors to include datasets and computer programs (source code); but only one instance of each was found in the 32 articles of vol. 4-7. The New York Journal of Mathematics (NYJM) mentions that articles contain links to supplementary material, related works, and comments; but the two volumes scanned for this study (vols 8 and 9) contained no examples of these supplementary materials. Sociological Research Online (SRO) 'encourages authors to make full use of the electronic media the journal publishes in, where possible attaching to articles, via hypertext links, written field notes and numerical datasets, oral data and other oral material, and visual materials including video data.' However, examples of this in the journal are extremely rare.

Most of the journals with embedded or linked data resources are found in the sciences. Links to data resources such as crystallographic, sequence and structure data are fairly common in highly specialized domains such as molec-

[13] See for example the first issue of Optics Express (OE) http://www.opticsexpress.org/abstract.cfm?URI=OPEX-1-1-2.

Table 5.14. Data resources in digital journals

- Databases / datasets (APS, CYBM, EARTHINT, GEOCHEM JAIR, JOP, LMS)
- Spectra, crystallographic data, molecular information (CECOMM, MOLEC, MOLVIS, PE)
- Image banks (DERMOJ)
- Mathematica Notebook files (EARTHINT)
- EXCEL worksheets (EPAA)
- Source code (JAIR, JASSS, LMS, PE)
- Integer sequences (JIS)
- Transcripts (MEDED)
- Music notation files (MTO)

ular chemistry. It should also be noted that the border between multimedia and data sources is diffuse, and that journals often do not make the distinction (e.g. in the example from SRO quoted above).

3.6 Revision

Almost no journals have specific provisions for revision of published articles, other than the inclusion of addenda and corrections in subsequent issues in the traditional manner. In general, therefore, it seems to be taken for granted that the published version is, in principle, final – as is also the case with printed journals. We found only very few examples where the journal offers more advanced options. Theory and Applications of Categories (TAC) states:

> 'The final accepted version of a paper is the form that will be in the archive; the author will not be able to make any changes, except:
> * errata and additions can be attached to the end and, where appropriate, footnote(s) may be added to the main text calling attention to these addendums;
> * references which mention pre-publication versions may be updated to final bibliographic form.'

Qualitative Report (QUALREP) promises that:

> 'Contributors to The Qualitative Report can also engage in a process known as Living Documents ... with their published works in the pages of the journal. In the Living Documents approach to writing, authors can nurture their published works and cultivate them as new developments arise

> within the scope of the paper, as hyperlinked resources located in the paper are updated or changed, and/or as the author's thinking evolves on the topic.'

However, in both cases we found no evidence in our sample that these features are used extensively by authors.

A number of journals do not allow additions and corrections to the original text, but do take into account that additions and revisions can be linked back to the original text:

> 'Internet Archaeology will not change content once it has been published, even if mistakes are discovered or if new data would render an interpretation obsolete. However we welcome subsequent addenda or 'new editions' of research published in the journal which can easily be linked to the original piece of work (and vice versa) thus building on its foundations.' (INTARCH)
>
> 'The article will be placed permanently in the EJDE; no further alteration will be allowed. However, authors may provide additional information by setting pointers to their web pages, or by attaching addenda/corrigenda to their articles.' (EJDE)
>
> 'Our papers are treated in the traditional manner, with no changes permitted after publication. But we also offer services that may be updated, such as links to reviews and related works, and archives of supporting materials (e.g., computer programs used for calculations in the paper), errata, and refereed commentary by the authors and others.' (NYJM)

There are only a few exceptions that acknowledge the the possibility of maintaining different versions as a specific property of digital publishing:

> 'JCSE is a Journal of Record, and papers may not be removed or altered (except for any editorial changes in format that are needed to accommodate new technologies) once publication has occurred. Minor corrections to papers should be made through the commentary process. However, when more significant changes are involved revised papers may be submitted, in which case the earlier paper and attached comments will be archived.' (JCSE)

In general, e-journals follow the prevailing practice that articles, once published, are final and are not open to revision by the author.

3.7 Response

In theory e-journals offer the possibility of a more active involvement of the reader, e.g. by attaching comments to articles and engaging in a dialogue with authors and other readers. In practice we found relatively few examples (12 journals, 6%) where journals provide for this type of interaction, and even less examples where interaction actually takes place. For instance, the Electronic Journal of Geotechnical Engineering (EJGE) announces the 'numerous advantages yet to be explored, such as live chat of the authors with discussers', but this has not materialized. EJGE also allows for extensive discussions:

> 'Discussions are open for submission from any W3G member or EJGE Subscriber that has questions or comments about the contents of a paper published by the EJGE. Any discussion submitted is subject to review and approval by the editor or an editorial board member.'

However, we found only one instance of this in all volumes (ca. 150 articles) from 1996 through 2004.[14]

[14] See http://www.ejge.com/2000/Ppr0003/disc/disc-tm.htm.

Comments

The Journal now offers the ability for readers to attach comments to specific paragraphs of a paper. There are currently three modes of viewing a paper:

- **View comments in a window** (the default mode) - in this mode the paper is displayed with a series of yellow 'notelets' down the left-hand edge of the screen. If the notelet is blank , then there are no comments at that point - you can make a comment by clicking on the notelet. If the notelet has writing on it , then there are comments at that point - you can read the comments, and make further comments if you wish, by clicking on the notelet.
- **Display comments inline** - in this mode all comments are displayed in the paper, after the paragraph that they are linked to. In this mode you can only add comments where one exists already, by clicking on the Add link.
- **Display for printing** - in this mode all comments are displayed in the paper, and the formatting of the page is optimised for printing. In order to get back to the normal display use the Back button on the browser, or click the JCSE logo at the top of the page, which will take you back to the JCSE Home Page.

An alerting system is also available to enable readers to be kept informed of the addition of comments to particular papers. To use this press the '**Watch this paper for new comments**' button.

Figure 5.4. Comments in Journal of Corrosion Science and Engineering (JCSE)

The Journal of Corrosion Science and Engineering (JCSE) uses various methods, including a system of 'notelets' attached to the article, for comments (fig. 5.4). However, the system does not seem to be used by the journal's readers.

Culture machine (CULTM) provides another example where user feedback functionality is explicitly announced:

> '... you will find at the end of each article published in the Culture Machine journal a hypertext button which will allow you to communicate your thoughts on the issues raised within that article to Culture Machine by e-mail. This 'Feedback' will then be published alongside the original article to which it refers, and readers will be able to access it either from the end of the relevant article, or directly via the graphic menu in the lower frame of the screen.'

But we found no examples, and this function does not seem to be working. European cells and materials (ECM) is another journal that explicitly solicits user comments for each article, but for the 49 articles published in vols. 1-8 we found only three articles with (extremely concise) comments (1 x 2, 2 x 1), of which one was 'excellent timely review'. Ecology and society (ECOSOC) has a function 'respond to article', but no examples of responses were found in vols 7 and 8 (2003). ECOSOC also has an 'open discussion forum', with only 4 postings in 2003. Other journals where we found little or no use of comment functions are: European Integration Online Papers (EIOP), Early Childhood Research & Practice (ECRP), Educational Policy Analysis Archives (EPAA), Journal of the Pancreas (JOP), New York Journal of Mathematics (NYJM), and Journal of Universal Computer Science (JUCS).

A special case is the Electronic Journal of Pathology and Histology (EJPH):

> 'The authors can choose to submit an article in two ways:
> 1. conventional (closed) version.This is the ordinary paper publication technique.
> 2. interactive (open) publication.This style of publication is offered to authors who are willing to accept an expansion by another team in the feature.'

Australian Humanities Review (AHR) is an exception in the volume of user response to articles published in the journal. It groups under the heading 'emuse' user comments on articles published in the journal. Out of some 165 articles published, 19 (11.5%) had comments attached with a total of 43 comments. The journal includes a comprehensive index of these comments.[15] The relatively high level of reader involvement here can be explained by the essayistic and sometimes provocative nature of the journal's contributions, and perhaps also a certain tradition in the field of cultural studies.

The Journal of Interactive Media in Education (JIME) offers a structured approach to user response by means of an outline window where readers can add comments on a per-chapter basis (fig. 5.5). However, these 'review discussion forums' are currently unavailable for the period 1996-2002. Later issues contain mostly invited comments (with occasional author's responses), but little or no sign of active reader involvement.

In some cases, readers are explicitly referred to other forums such as discussion lists:

> 'The moderated listserv ACMTNet serves as the vehicle for posting commentary, discussion or questions regarding materials published in [the journal].' (IJMT)

So with very few exceptions, e-journals do not succeed in involving the reader in an interactive dialogue or even in soliciting comments that could be attached to the original article.

3.8 Customization

There is very little evidence that authors, editors or publishers have in mind a user who might want to adapt the structure or presentation of an article to his or her individual requirements or preferences, nor do they envisage a variety of users with different characteristics to which the journal's output might be adapted. In the majority of cases, journals apply strict formatting rules. They often attempt to emulate the typographic characteristics of print journals, and often also use formats such as PDF that are aimed at preserving the typography of the original publication.[16] Only one journal, the Internet Journal of Chemistry (IJC), suggests that it provides mechanisms that allow the user to adapt the presentation of published materials:

[15] See http://www.lib.latrobe.edu.au/AHR/emuse/home.html and http://www.lib.latrobe.edu.au/AHR/emuse/index.html.

[16] See also section 3.14.

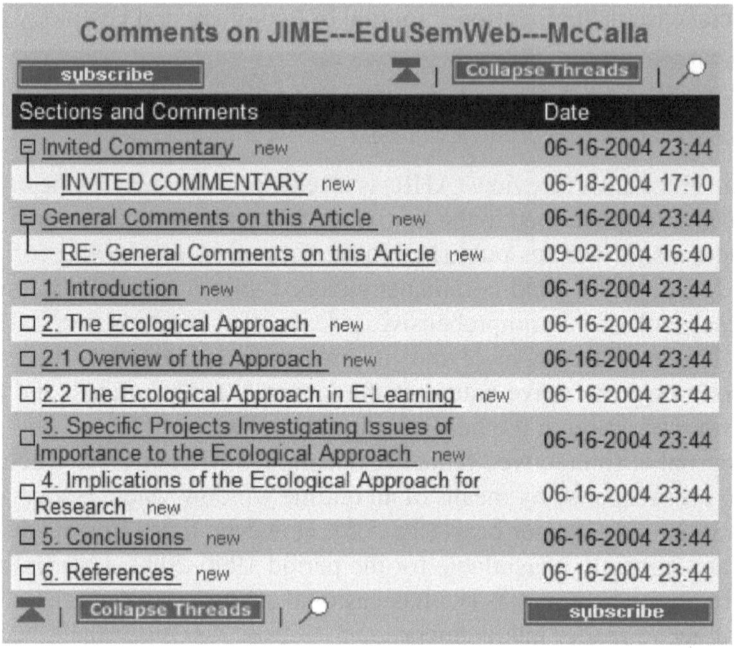

Figure 5.5. Comments in JIME

> We define interactive tools in a very broad sense, encompassing all tools that allow the reader to manipulate data and change the nature of the presentation. We strongly encourage authors to incorporate these tools to the fullest extent. ... We also have made available a tutorial on creating enhanced pages which further explains the use of these interactive tools.[17]

However, what is meant here is something slightly different: the use of 3-D models and virtual reality modeling that allows the user to manipulate visual objects, where their use would be prohibited in a print journal. But IJC does contain a set of interesting customization features that control the multimedia functionality of the journal.[18]

Apart from the possibility to adapt the screen presentation of html pages in a browser window (e.g. window size, font type and size), e-journals offer the reader little or no ways to adapt the structure and presentation of articles to their own needs. It is interesting that in one of the few examples where a journal seems to be aware of differing user requirements with respect to presentation, the issue is related to print rather than screen output:

> 'The restriction to LaTeX format is justified by the additional flexibility gained. The most basic example of this flexibility is that it allows readers in Europe to use their wide paper format when

[17] See http://hackberry.chem.trinity.edu/IJC/Text/index.html.
[18] http://www.ijc.com/pref.papers.html.

Table 5.15. Customization features in Internet Journal of Chemistry (IJC)

- Greek characters visualization (GIF-images or Symbol Font)
- Interactive chemical structures (standard GIF/JPEG image, link to datafile, as embedded object, Java applet)
- Interactive features default size and style
- Use of multimedia resource icons
- On-th-fly units conversion (accessibility and presentation)
- Footnotes display (bottom of page or independent window)
- Units enforcement (original units or automatic conversion)
- Notation (simple or scientific)

printing an article while simultaneously allowing the readers in the USA to use a smaller paper format.' (JFLP)

Only recently (with its issue of January 2005), the Journal of Computer-Mediated Communication (JCMC) has introduced the choice of smaller or larger text, and justified or unjustified margins when reading JCMC articles online.

In our sample we found no evidence of journals that use characteristics of the reader to adapt content or presentation. At the aggregate level, some customization is offered by a limited number of journals in the form of 'personal folders' (JHEP) or 'MyArticles' (ARLO) that allow the user to store selected articles for subsequent use. Another form of personalization is the alerting service that sends an e-mail to subscribers when new articles appear in the journal. We found examples of this in 38% of the journals in our sample.

3.9 External hyperlinks

Hyperlinks are perhaps the most distinctive property that differentiates digital documents from print. They function either as internal navigational devices (see section 3.11) or as external pointers to other resources (including publications, organizational websites and personal e-mail addresses) available through the network. The point about hyperlinks is that they are not mere textual references, but that they are *functional,* i.e. when they are activated (e.g. by clicking on them with a pointing device) they perform the function of replacing the representation of the current document with a representation of the referenced document.

Table 5.16. Links in html journals

Type	Number of journals	% of all journals
Activated	85	68%
None or not activated	40	32%
Total	125	100%

It is reasonable to expect that a journal that sets out to exploit the digital format will at least include hyperlinks in the aforementioned sense.[19] The most common publication formats used for e-journals (html and pdf) allow for the inclusion of hyperlinks, but postscript (used in 11% of our sample) does not normally support hyperlinks.[20] What is interesting, though, is that in about 40% of the e-journals published in html and/or pdf external hyperlinks are either absent or not activated, i.e. are not presented as active hyperlinks but only as a textual representation.[21] It should be noted, however, that the non-use of external hyperlinks can sometimes be explained by the lack of networked resources within the field covered by the journal (at least during the period covered in our study). Also, a check on recent issues shows that the use of hyperlinks is increasing. Tables 5.16 and 5.17 provide the details for both html and pdf. As might be expected, hyperlinks are much less common in pdf than in html. It is not unusual to find that when a journal is published in both formats, the hyperlinks in the pdf version are not activated.

Some journals actively encourage the use of hyperlinks and offer support to authors with the input:

> 'Authors of contributions accepted for publication in the journal are encouraged to explore and to implement hyperlinks to books, articles, and other sites of interest (including such as publishers' web sites) and input the name of the item with the URL of it in parentheses in the main text as well as in the works cited of the paper, as required (CLCWeb activates the hyperlink when uploading the text to the server).' (CLCWEB)

> '[The journal] encourages authors to make full use of the electronic media the journal publishes in, where possible attaching to articles, via hypertext links, written field notes and numerical datasets, oral data and other oral material, and visual materials including video data.' (SRO)

[19]Of course, print journals also contain references to electronic literature in the form of url's, and these are increasingly activated (often by the journal editors) as hyperlinks in the digital versions of these journals. Most of these hyperlinks reference other digital versions of formal scientific publications (Wouters and Vries 2004).

[20]If the source of a postscript file (e.g. the dvi-output of latex converted to postscript through dvips) contains hyperlinks, subsequent conversion to pdf (e.g. through ps2pdf) will retain the hyperlinks, which can be rendered by a pdf-reader such as acrobat. Pdf is essentially a version of encoded compressed postscript with a full hyperlink overlay.

[21]This figure is approximate: in a number of cases practice differs between articles, suggesting that editors do not always find it necessary to activate network addresses where the author has not done so in the input file.

Table 5.17. Links in pdf journals

Type	Number	Percentage
Activated	19	38%
None or not activated	31	62%
Total	50	100%

NB: calculated for journals published in pdf only (no html)

There are also, however, cases where hyperlinks are not welcomed by the journal's editors:

> 'In general, we prefer not to link EJC texts with web sites external to CIOS resources. This is because such sites tend to become dead within a relatively short period of time. Please do not link URLs in the citation list to existing web sites. It is permissible to use URLs as part of your citation, following conventional style sheet requirements, but avoid constructing links to external resources.' (EJC)
>
> 'Use of external links from articles is possible, but the authors are must ensure that the pointed-to material will be on that location for at least 5 years after the paper is published. Links must, however, be useful on printed copy as well.' (EJITC)
>
> 'Accepted papers may contain up to 5 hyper links to related web resources.' (HYLE)

Hyperlinks are sometimes used as the mechanism for providing added functionality, such as links to works published by the same author (ELAW) or to articles on the same subject (BQUEST). In a limited number of cases the references provide links to 'digital repositories' of published papers such as the *ACM Digital Library*[22] (ACMJEA), *CiteSeer*[23] (JAIR), *MathSciNet*[24] (CGD), *PubMed*[25] (IJIC), or the *Clearinghouse on Early Education and Parenting*[26] (ECRP).

3.10 Functionality

Our sample provided few examples of 'functional documents' (i.e. articles with embedded functionality) along the lines described in Chapter 4.6. A notable exception is Internet Archaeology where some articles have embedded database functionality. For instance, an article in the first issue provides examples of access to underlying databases using image maps and time-lines as the query interface.[27] On the whole, however, the use of embedded functionality is extremely limited. The Journal of Academic Leadership announced a

[22] http://portal.acm.org/dl.cfm.
[23] http://citeseer.ist.psu.edu/.
[24] http://www.ams.org/mathscinet.
[25] http://www.pubmedcentral.nih.gov/.
[26] http://ceep.crc.uiuc.edu/.
[27] Peacey 1996.

Table 5.18. E-journal functions

- General search functions (EGJ, EJANZ, EJCOMB, ELANT, ELAW, ETNA, IMEJ, INTARTCH, INTERSTAT, JAHC, JAIR, JASS, JFLP, JILT, JMEM, JSE, MOLVIS, MRS, NEXUS, PSYCHO, SRO, TRANS, WJCLI)
- Advanced search functions: search related works (BPO); find similar (JMR), similar docs (JUCS)
- Links to works by same author (ELAW)
- Links to articles on same subject (BQUEST)
- Find citations (BQEST); download citations (ARLO)
- MyArticles (ARLO); personal folders (JHEP)
- On-the-fly units conversion (IJC)
- User lay-out customization (IJC)
- Convert to pdf (JMR)
- Translate to other language (ACL)

function for translating articles into other languages, but the journal's content is temporarily unavailable.

We did, however, find various examples of user-oriented functions at the aggregate level of the journal, some of which, such as response functions, have already been discussed (table 5.18). The main functionality found in our sample is the search function. This includes textual search within a single article, keyword searching of the journal space (journals/issues/articles) and advanced searches for finding similar/related articles.

3.11 Navigation

One of the frequently heard objections against digital documents is the difficulty users have with navigating digital information spaces.[28] In order to help the user, digital documents can adopt a variety of navigational devices that

[28]Dillon et al., 1990 and Edwards and Hardman 1999. See also O'Hara and Sellen 1997.

provide content lists (typically tables of contents, figures etc.), forward and backward movement, bookmarking, content visualizations (e.g. maps), etc. We have looked to find such navigational devices in our sample of e-journals.

It is useful to distinguish between internal navigation at the level of the individual article, and links to other elements of the journal at the aggregate level. Examples of internal links are hyperlinked tables of contents and similar devices (either at the top of the document or as a separate frame or box), and links to and/or from structural text elements such as footnotes, references, tables and figures. Examples of aggregate links are links to the table of contents of the journal issue, and links to other articles within the journal (e.g. links to papers by the same author (ELAW) or to articles on a similar subject (BQUEST)).

Table 5.19. Types of navigation

- Journal structure (e.g. volume / issue / article)
- Content structure (e.g. chapters, sections)
- Content elements (e.g. tables, figures, equations)
- Notes (endnotes or footnotes)
- References
- Movement (e.g. forward / backward / beginning / end)

The navigational devices found in e-journals are listed in table 5.19. These devices either navigate the journal as a document space (i.e. a set of volumes, issues and articles), or the single article. Navigation within an article can be based on an outline of the content structure, on specific content elements (e.g. tables) and on additional elements such as notes and references.

Most navigational devices found in our sample are either hyperlink lists (e.g. in the form of tables, frames or panels) or in-text hyperlink references. Most in-text hyperlinks are used for forward referencing, but sometimes we also found backward referencing, e.g. from notes or bibliographic references back to the link in the text.

The Journal of Inequalities in Pure and Applied Mathematics (JIPAM), using pdf as its publishing format, has a panel on the right side of the screen that allows the user to navigate through the article and rapidly access a hyperlinked content outline (fig. 5.6). Since the panel is linked to a pdf-version of the article, no navigation at the aggregate level of the journal is possible.

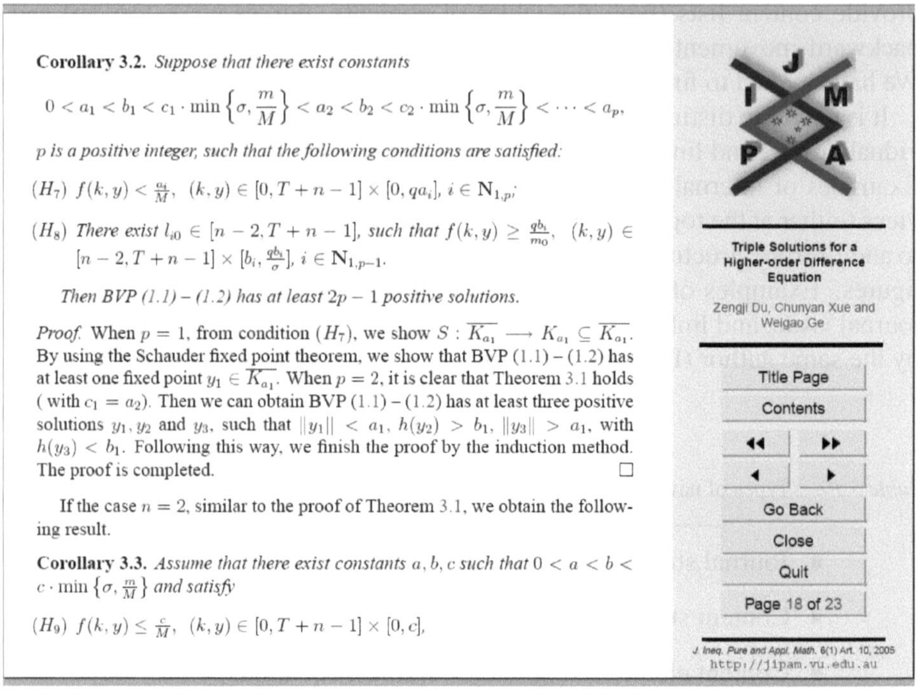

Corollary 3.2. *Suppose that there exist constants*

$$0 < a_1 < b_1 < c_1 \cdot \min\left\{\sigma, \frac{m}{M}\right\} < a_2 < b_2 < c_2 \cdot \min\left\{\sigma, \frac{m}{M}\right\} < \cdots < a_p,$$

p is a positive integer, such that the following conditions are satisfied:

(H_7) $f(k, y) < \frac{a_i}{M}$, $(k, y) \in [0, T+n-1] \times [0, qa_i]$, $i \in \mathbf{N}_{1,p}$;

(H_8) *There exist* $l_{i0} \in [n-2, T+n-1]$, *such that* $f(k, y) \geq \frac{qb_i}{m_0}$, $(k, y) \in [n-2, T+n-1] \times [b_i, \frac{qb_i}{\sigma}]$, $i \in \mathbf{N}_{1,p-1}$.

Then BVP (1.1) – (1.2) has at least $2p-1$ *positive solutions.*

Proof. When $p = 1$, from condition (H_7), we show $S : \overline{K_{a_1}} \longrightarrow K_{a_1} \subseteq \overline{K_{a_1}}$. By using the Schauder fixed point theorem, we show that BVP (1.1) – (1.2) has at least one fixed point $y_1 \in \overline{K_{a_1}}$. When $p = 2$, it is clear that Theorem 3.1 holds (with $c_1 = a_2$). Then we can obtain BVP (1.1) – (1.2) has at least three positive solutions y_1, y_2 and y_3, such that $\|y_1\| < a_1$, $h(y_2) > b_1$, $\|y_3\| > a_1$, with $h(y_3) < b_1$. Following this way, we finish the proof by the induction method. The proof is completed. □

If the case $n = 2$, similar to the proof of Theorem 3.1, we obtain the following result.

Corollary 3.3. *Assume that there exist constants* a, b, c *such that* $0 < a < b < c \cdot \min\{\sigma, \frac{m}{M}\}$ *and satisfy*

(H_9) $f(k, y) \leq \frac{c}{M}$, $(k, y) \in [0, T+n-1] \times [0, c]$,

Figure 5.6. Navigational devices in Journal of Inequalities in Pure and Applied Mathematics (JIPAM)

Living Reviews in Relativity (LRR) offers the user extended navigation functionality with navigational links, reference links and a navigational sidebar (fig. 5.7). The sidebar (that can also be represented as a separate window) gives access to a number of functions: besides a structured outline of the article several aggregate functions such as a 'compendium' (in fact the homepage of the journal), an article contents list of the journal, a search function (for searching in LRR), the article history (listing revisions etc.), the article's bibliographic data, various download options (including the reference list formatted in bibtex) and a help function. LRR makes a conscious effort to create a hypertext journal rather than an html-rendering of a traditional print journal:

> 'One of the main features of Living Reviews is the provision of a hypertext viewing environment. With a print document, a viewer tends to read an author's ideas in a preordained manner. With a hypertext document, a viewer navigates information by following links on an "as-needed" or an "as-desired" basis. Thus, the path one follows through a hypertext document can result in an individual experience: self-selected and thus self-directed. Given this dynamic quality of hypertext, we expect Living Reviews to be used rather than just read. The awareness that we have users, not just readers, is a major principle guiding the design of Living Reviews.
>
> A user working with electronic documents needs information to be presented in a way that is easy to assimilate from a computer screen. Most users access Living Reviews from their working

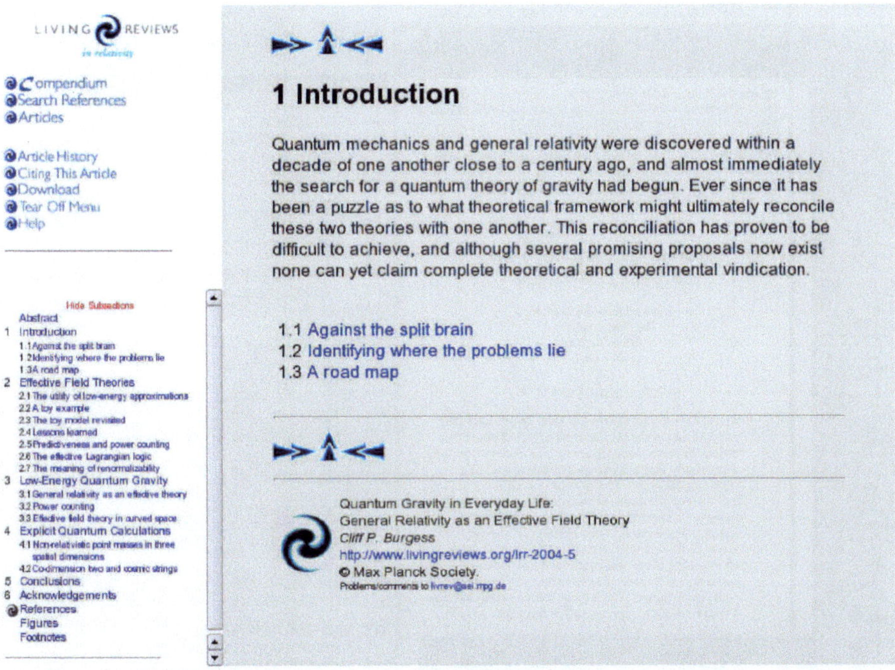

Figure 5.7. Navigational devices in Living Reviews in Relativity (LRR)

environment, with all its interruptions and distractions. They often suspend their own reading of an article to follow up a link to an on-line resource, to another section of the article, to a reference, or to an index. These jumps may lead to further links. Such a style of use puts certain requirements on the journal's presentation and content. Deep and complex arguments, detailed proofs of theorems, and long mathematical derivations are probably better communicated on paper. Living Reviews is, therefore, not a textbook or a monograph. Rather, it offers surveys of recent work, evaluations of the importance and interconnections of results, summaries of important results, entry points into the essential literature, assessments of where new progress is needed, access to web sites and other useful electronic contacts, and databases of the recent literature.'[29]

The Journal of Interactive Media in Education (JIME) has in its html-version a hyperlinked table of contents per article in a separate frame (fig. 5.8); the figure also shows the frame for commentary discussed in section 3.7). The journal is also published in a pdf-version that emulates a traditional print journal in its typography (fig. 5.9), creating a striking difference with its html counterpart.

[29] http://relativity.livingreviews.org/Info/AboutLR/concept.html.

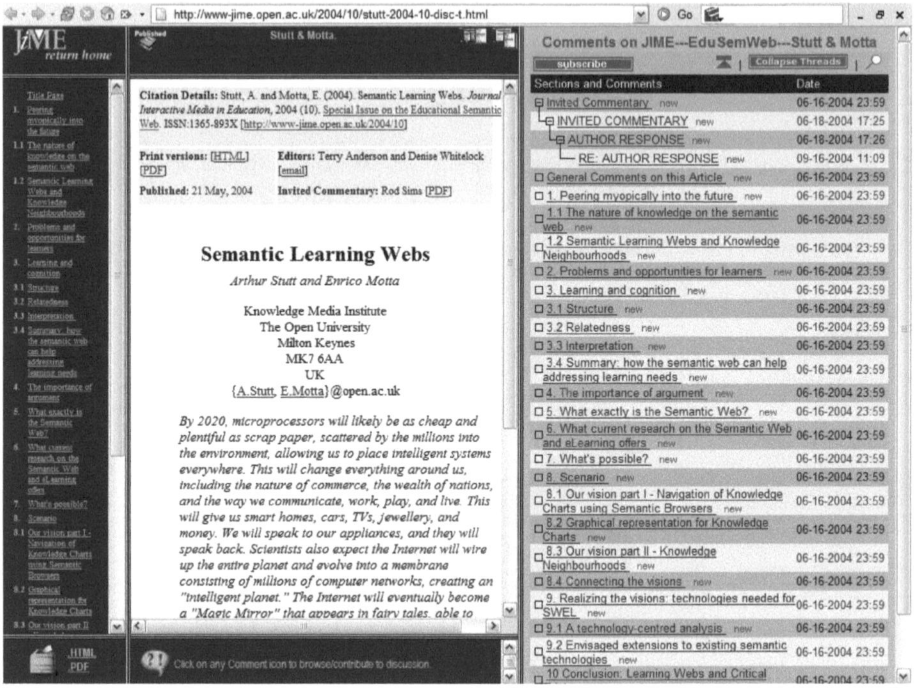

Figure 5.8. Navigational devices in Journal of Interactive Media in Education (JIME)

As with external hyperlinks, the extent to which hypertext navigational devices were used in our sample was less than one might expect. Some 105 (56%) journals employed navigation within articles (although only 12.5% navigate the article's structure through hyperlinked tables of contents), and in 25 (13%) cases we found journal structure navigation at the aggregate level. Surprisingly, of the 125 journals with html output format, 35 (28%) do *not* use any kind of navigational links within the article. There is also a great variety (and often inconsistency) in the types of navigational functions provided. For instance, many journals provide links to (as well as sometimes *from*) footnotes, but not to references. Of the 79 journals in our sample with hyperlinks to notes and/or references, 20 (25%) contain links to notes only, not to references. Navigational devices that are included in the html-version are often not included in the pdf version, even though pdf does allow for hyperlinks and hyperlinked content tables. Apparently, editors regard html as a format for on-screen reading and pdf for printing.[30]

[30] An example of the different approaches to html and pdf is the Journal of Interactive Media in Education (JIME), see sections 5.8 and 5.9.

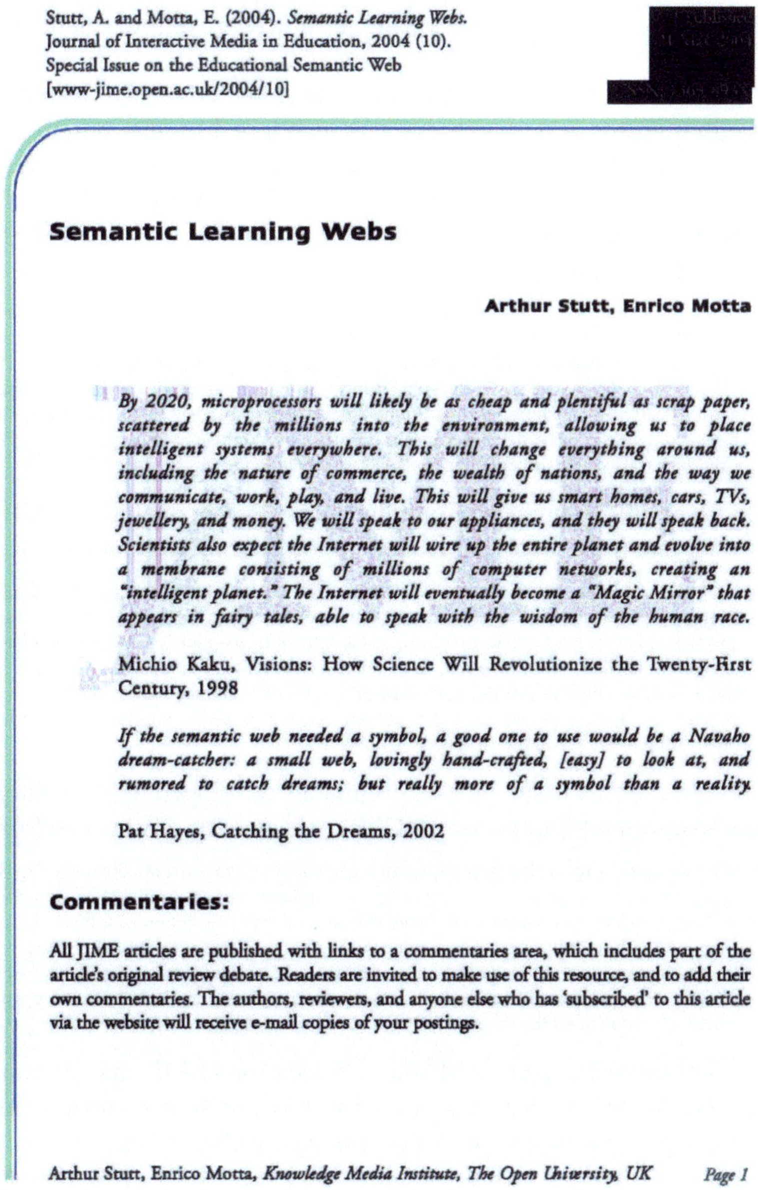

Stutt, A. and Motta, E. (2004). *Semantic Learning Webs.*
Journal of Interactive Media in Education, 2004 (10).
Special Issue on the Educational Semantic Web
[www-jime.open.ac.uk/2004/10]

Semantic Learning Webs

Arthur Stutt, Enrico Motta

By 2020, microprocessors will likely be as cheap and plentiful as scrap paper, scattered by the millions into the environment, allowing us to place intelligent systems everywhere. This will change everything around us, including the nature of commerce, the wealth of nations, and the way we communicate, work, play, and live. This will give us smart homes, cars, TVs, jewellery, and money. We will speak to our appliances, and they will speak back. Scientists also expect the Internet will wire up the entire planet and evolve into a membrane consisting of millions of computer networks, creating an "intelligent planet." The Internet will eventually become a "Magic Mirror" that appears in fairy tales, able to speak with the wisdom of the human race.

Michio Kaku, Visions: How Science Will Revolutionize the Twenty-First Century, 1998

If the semantic web needed a symbol, a good one to use would be a Navaho dream-catcher: a small web, lovingly hand-crafted, [easy] to look at, and rumored to catch dreams; but really more of a symbol than a reality.

Pat Hayes, Catching the Dreams, 2002

Commentaries:

All JIME articles are published with links to a commentaries area, which includes part of the article's original review debate. Readers are invited to make use of this resource, and to add their own commentaries. The authors, reviewers, and anyone else who has 'subscribed' to this article via the website will receive e-mail copies of your postings.

Arthur Stutt, Enrico Motta, *Knowledge Media Institute, The Open University, UK* *Page 1*

Figure 5.9. PDF-version of Journal of Interactive Media in Education (JIME)

3.12 Peer review

One of the expected outcomes of the so-called 'Gutenberg revolution' in scientific publishing that is often mentioned is the transformation of peer review as the main system of quality assurance in the scientific communication system. Most journals in our sample, however, follow standard peer review practices.

We did find some 15 (8%) cases where the advantage of e-publishing for speeding up the review process was mentioned by the journal's editors. We also found 6 (3%) cases of ex-post peer review, i.e. where paper were subjected to some form of review process *after* having been published in the journal. The Electronic Journal of Cognitive and Brain Sciences (EJCBS) describes its procedures as follows:

> 'Each document may have two status: (1) "Submitted" and (2) "Accepted" status. From the moment of submission the document will be public and open for "commentaries" for a [month]. Then the status of the paper is finalized based on the evaluations submitted by the readers. The final status can be an "accepted" or a "rejected" status. At the second phase the accepted document will automatically be transferred to the "Archive of Accepted Papers" (Table of Content [accepted]). Rejected papers will be deleted from the "Temporary Archive" (Table of Content [Submitted]). Commentaries will be accepted or rejected together with the target paper.' (EJCBS)

The Electronic Journal of Communicative Psychoanalysis (EJCP) has developed a system for involving readers in the peer review process:

> 'Before considering an articles for publication, The Editor will request selected Readers to write Peer Reviews which will be offered to the authors for thoughtful consideration. The comments and the names of the Reviewers will be made known to the authors of corresponding articles to encourage fairness of critique and to discourage anonymous influence. The main aim of Peer Reviews is to cultivate the usage of scholarly style of argument and to offer comments on the content. Reviewers are encouraged to submit discussions of the articles they have reviewed for publication under the rubric Response Articles.'

A review process based on user comments developed by the Journal of Corrosion Science and Engineering (JCSE), was not entirely successful:

> 'From Volume 4, the review process is moving towards a more conventional invited review process, as the commentary process has not yet become sufficiently widely used. On submission, papers will be published in Preprint form. This will use a direct production of a PDF file from the text supplied, with no editorial modification (although the Editor reserves the right to reject papers for any reason - this will normally only be done in the case of papers that are illegal, indecent or trivial). Two referees will be invited to review the Preprints (and it will also be possible for any reader to comment on the Preprints).'

The Journal of Interactive Media in Education (JIME) has developed a novel submission system that includes a form of public peer review before formal publication combined with ex-post reviewing (fig. 5.10)[31]

The British Medical Journal, although a print publication (and therefore not included in our sample) maintains a website that may in future be used to organize public peer review:[32]

> 'The BMJ has until now used a closed system of peer review. ... Now we plan to let authors know the identity of reviewers. Soon we are likely to open up the whole system so that anybody

[31] http://www-jime.open.ac.uk/about.html. See also Sumner and Shum 1996; Shum and Sumner 2001.

[32] Note that this announcement is from 1999; public peer review still has not been introduced.

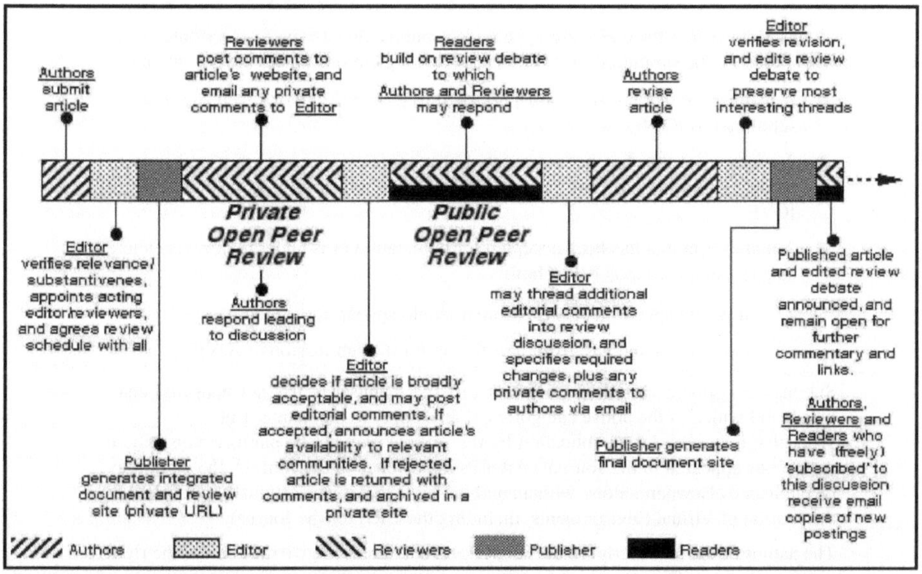

Figure 5.10. Peer review process in Journal of Interactive Media in Education (JIME)

> interested can see the whole process on the world wide web. From this week ... the BMJ will identify to authors the names of those who have reviewed their papers, including the names of our in house editorial and statistical advisers. But we expect to go further, researching as we go. Soon we will probably start to list reviewers at the end of articles. Then we may move to a system where authors and readers can watch the peer review system on the world wide web as it happens and contribute their comments. Peer review will become increasingly a scientific discourse rather than a summary judgment. Through such openness we will hope to show that peer review by journals does add value to the scientific process and that we will thus have a place in an electronic world where authors can potentially go straight to readers.' [33]

3.13 Copyright

The approach to copyright does not seem to be very much different from that in print journals. Copyright information for authors was provided for 117 (63%) journals. In 50% of these cases (58 journals) authors are required to transfer copyright to the publisher. In the other 50% (59 journals) authors retain their copyright. However, when authors retain their copyright, they are generally required to allow the journal the right of first publication and to include a reference to the first publication in subsequent publications (fig. 5.11). In some cases journals explicitly disallow prior or parallel publication – we found this type of restriction in 58 (33%) journals. But we expect that most editors presume that manuscripts presented to the journal are original and have not been submitted elsewhere. Only very few journals explicitly allow prior

[33] Smith 1999b.

Manuscripts which have already appeared in publication (print or electronic) elsewhere may not be submitted to JOVE. The following circumstances are acceptable:

- A manuscript that has been self-published (e.g. on the author's web site) may be submitted to JOVE.
- A manuscript that has been presented at a conference as a poster, oral presentation, workshop, etc. but has not appeared in publication may be submitted to JOVE.
- A manuscript that has been accepted for publication or is under review elsewhere may be submitted to JOVE if both:
  - it would appear in JOVE before it would appear in the other venue
  - the other venue has given permission for the submission to JOVE

Submissions should include a statement that the manuscript has not appeared elsewhere and which of the above categories, if any, apply to the manuscript.
Work that is accepted for publication by the Journal may not be published elsewhere until it has appeared in the Journal. After its appearance in the Journal, the work may be published elsewhere, along with an author's note that it was originally published in the Journal of Virtual Environments, including the URL for the Journal.

The author of the manuscript retains copyright, but agrees to give the Journal the right of first publication.

Figure 5.11. Copyright policy of the Journal of Virtual Environments (JVE)

or parallel publication. Sometimes only the copyright of the digital version is transfered, while print copyright is retained by the author (CHEMJI). We also found one instance (INFRES) that explicitly refers to the Creative Commons License.[34]

3.14 Editorial policies

In the previous paragraphs we have already included a number of examples that suggest that journal editors influence – and often restrict – the extent to which authors exploit the digital format in their writings. In this section we provide a more detailed analysis of our findings in this area.

Style and structure

As with print journals, most e-journals present authors with strict rules about the style of the publication. In most cases (56%) the journal prescribes its own editorial style, but in other cases (20%) the author is referred to existing style manual (American Institute of Physics (AIP), American Psychological Association (APA), Modern Language Association (MLA) and others). In almost

[34] See http://creativecommons.org/licenses/by-nd-nc/1.0/legalcode.

Table 5.20. Editorial style rules

Style element	Number of journals	% of all journals
Length	46	25
Spacing	36	19
Margins	31	17
Page size	31	17
Font	48	26
Headings	27	15
Structure	13	7

25% of the journals no style rules are mentioned explicitly, although these are often cases where the journal's editors apply formatting to the author's input according to some unspecified set of style rules. The general impression is that, as with print journals, electronic journals tend to have their own, distinct presentation style. Almost 15% of the journals supply template files (e.g. rtf or Word templates, or latex style files).

An overview of the extent to which editors provide strict instructions for textual elements is given in Table 5.20. The level of detail of these requirements is often somewhat surprising, in view of the fact that the digital medium offers far greater flexibility for authors to determine the visual presentation of their texts, and far less limitations e.g. on article length. Our findings suggest that e-journal editors still regard the electronic publication as a modern version of print as far as presentation style is concerned.

Only occasionally do we find a more liberal attitude:

> 'The style of writing for contributions to The Qualitative Report is also a matter of particularity for authors. Given the subtle, and, sometimes, not so subtle ways in which a style guide can shape the writing choices made by an author, contributors may take greater latitude in making stylistic choices in their manuscripts. The Publication Manual of the American Psychological Association (5th ed.) is used as a guide for contributors to The Qualitative Report only when it comes to the format of references and citations within the text. Otherwise, style is a matter of choice on the parts of authors.' (QUALREP)

But sometimes the structure of articles is strictly enforced for a specific reason:

> 'These sections must imperatively be used since they will be required for automatic processing of the manuscript. Titles of sections must be contained in brackets.' (EJO)

In general, however, style and structure seem to be governed by the traditional rules prevalent in a specific domain. This is particularly illustrated by the use of style guides such as the APA and MLA styles, or derivatives thereof.

Limits on length etc.

Editors of e-journals are often slightly confused about the volume of information that can sensibly be published in this way. Although some recognize that there are no technical or economic limitations to the size of articles, they do express concern about the capability and willingness of readers to absorb large quantities of information, especially when reading from the screen:

> 'There is no page limit for articles published in Palaeontologia Electronica' (PE)
>
> 'Because online publication is not bound by the costs associated with print-based publication, articles can be of any length.' (EJANZ)
>
> 'Because space is not an issue in web publication, authors are encouraged to include additional reference material, such as complete texts of archival documents, glossaries, biographical notes, etc., in one or more appendices to the article, to be linked to specific words or short phrases within the article or its notes.' (JSCM)
>
> 'Shorter sentences and paragraphs are best suited to electronic publications. Readers need text that is concise and useful rather than wordy and general; therefore, writing to express rather than impress will best promote your ideas.' (FM)
>
> 'There is no limit on the length or size of papers, although the Editor reserves the right to ask for a reduction in size of unreasonably large papers. Prospective authors should, however, appreciate that online readers will typically not read to the end of a very long paper, and conciseness is even more important for JCSE than for paper publications.' (JCSE)
>
> 'As an electronic-only journal JoDI does not prescribe new rules for the well-established structure of academic papers, but provides the framework and capability for you to enhance and add to your presentation as you wish. There is no limit to the length of papers or to the amount of data you can include, but no reader has unlimited time. Available evidence suggests, in fact, that readers of Web pages are less tolerant of wastefulness. So, it is vital that your paper is concisely written.' (JODI)
>
> '... the nature of electronic journals suggests that a reasonable length (generally no more than 5000 words) is appropriate.' (NHAE)

Restrictions on the use of various textual elements such as footnotes etc. are surprisingly common. Current Research in Social Psychology (CRISP) instructs its authors that endnotes/footnotes, figures and the use of symbols and other special characters should be avoided. Other examples abound:

> 'The use of codes affecting the layout of the paper should be avoided altogether. Endnotes should not be used. Footnotes should be used sparingly: in general, to give sources of direct quotations, references to authorities, and evidence relied on. Do not use any footnote-options.' (ARSDIS)
>
> 'Authors are asked to strictly limit their use of underlining and bolding. Please keep tables and graphs to a minimum. All articles should have page numbering, and there should be no section breaks.' (ASQ)
>
> 'Only indispensable graphs and diagrams are to be presented, and tables should be in three-line format. No more than six graphs are allowed in an article, either in color or in black and white' (CHEMJI)
>
> 'Tables and figures, a maximum of three each, must be included within the text of the paper.' (EGJ)
>
> 'Authors are requested not to submit e-scripts containing more than 10 figures or photographs. It must be kept in mind that transfer of these files by the internet may be time-consuming.' (EJO)
>
> 'Footnotes and endnotes are not accepted, all such information should be included in the text' (EDTS); 'Please make only scarce use of' footnotes' (EIOP); 'We encourage authors to use footnotes sparingly, since they may be difficult to read online' (JAIR); 'Please do not use footnotes or endnotes' (SCOPE); 'The use of footnotes should be minimized' (SWJPAM); 'Notes should be endnotes, and kept to a minimum' (TESL).

Formatting/typography

As already mentioned, in many cases the digital content is created through the editorial process from more or less standard, unformatted material supplied by the author (unless, of course, camera-ready copy – usually submitted as postscript/pdf – is required). Electronic Antiquity ELANT tells its authors not to be concerned about specific formatting: 'If the submission is accepted, the editors will work with the author on formatting.' Some journals are more explicit in their discouragement of explicit formatting by the author:

> 'Papers should be submitted as 'clean' as possible, that is to say that the standard settings of the word-processing software used should in no way whatsoever be changed.' (ARSDIS)
>
> 'No attempt should be made to arrange the text into columns, or apply other formatting.' (ECM)
>
> 'Authors are requested to keep the text as plain as possible if Word or WordPerfect format is used; i.e., use a minimum of layout, avoid styles, etc., in order to facilitate the conversion process.' (EJCL)
>
> 'The main text should be sent in as a non-formatted Word file (single spaced, flush left). In other words, you should not use a hyphenation program, a footnote program or any special printing formats. You may send an additional Word or Adobe file or a printed copy of the text in which one can see the layout as you would like to have it (structure, tables, and so on).' (FQS)
>
> 'Note that Information Research uses a style sheet, which fixes the style of paragraphs and headings; therefore, do not use any style features (such as font size or colour) in your own HTML code' (INFRES)
>
> 'Please do not mark up your paper in HTML. You can send it either as ASCII text or as an attached e-mail file. We will mark it up and format it in a standard style.' (ISTL)
>
> 'Do not use formatting such as Word's 'Heading' or 'Style Sheets' as all submissions will be reformatted.' (JPPS)

Other journals, however, are more liberal:

> 'The layout of the paper is left to the authors. We do not impose any particular style, nor do we provide any macros to make the paper. The only thing to keep in mind is that your paper will be printed both in A4 format (29.8cm x 21.0cm) and in Legal format (27.9 x 21.6cm), so do not make the pages too wide or too tall.' (MPEJ)

Digitality

Surprisingly, we found explicit statements about the digital nature of the journal, its consequences for authors and guidelines for its utilization in only just over a quarter of the journals (27%). These editorial statements tend to be framed in positive terms, emphasizing the added value and new possibilities, especially in terms of multimedia representation, hyperlinks and inclusion of large data quantities:

> 'Our mission is to disseminate scientific information worldwide, taking full advantage of the electronic publication medium by offering 3D graphics, video, interactive figures, ancillary databases, and sound.' (APS)
>
> 'The journal's mission is to disseminate scientific and technical information on drug product design, development, evaluation and processing to the global pharmaceutical research community, taking full advantage of web-based publishing by presenting innovative text with 3D graphics, interactive figures and databases, video and audio files. All authors are encouraged to take full advantage of the Web-only capabilities of online publishing, including 3-D, video, and interactive graphics.' (APST)
>
> 'Currents in Electronic Literacy is a peer-reviewed journal that encourages submissions that take advantage of the hypertext and multimedia possibilities afforded by our World Wide Web publication format, as well as articles concerning the use of emergent electronic technologies. To this end, we gladly accept articles with graphics, sound, and hyperlinks submitted as HTML documents.' (CURREL)

> 'A driving force behind the creation of this journal as an electronic journal was the desire to free authors of the constraints of the static printed page in the presentation of their research results. Dynamic and interactive components have already appeared as an integral part of articles published in Earth Interactions and this can only happen effectively if authors do not feel the need to also provide printable forms of this material in addition to the dynamic forms.' (EARTHINT)
>
> 'Web articles need not be limited to the conference presentation model. Hypertext links (non-sequential or multiply-branching texts) free the author to experiment with non-linear and user-directed pages. Authors are invited to experiment with this new medium. A Web publication can simulate a conference presentation. Include audio, graphics, or video illustrations just as you would in a conference paper.' (EOL)
>
> 'Authors are encouraged to take advantage of the expressive possibilities afforded by JCMC's multimodal, web-based format. Articles may contain any combination of text, tables, graphics, animation, or audio components. Innovative forms of expressing research, and/or linking members of the scientific community, are welcome.' (JCMC)
>
> 'Submissions should take advantage of the multimedia capabilities of the World Wide Web, that is use audio, graphics, or video and preferably integrate text and multimedia. Text-only manuscripts will be considered inappropriate. A Web article may be conceived in many different ways, for instance as one text arranged in sections and illustrated with images, audio and video examples; or as non-sequential or multiply-branching texts dealing with different aspects of the chosen topic. M&A and Ethnomusicology Online offer examples of multimedia Web style.' (MA)
>
> 'Unlike many print-based journals, Palaeontologia Electronica is highly graphical in both its format and content. Authors are encouraged to make use of color in their figures and tables and to include high-resolution digital images as illustrations. Moreover, Palaeontologia Electronica encourages active experimentation with animation, 2D and 3D modeling of morphologies, on-line access to databases, and the creation of on-line data analysis tools.' (PE)

These editorial comments invariably stress the difference between print and electronic, the advantages of the digital communication, and encourage authors to benefit from the new medium and its freedom by experimenting with new modes of expression and representation. However, some editors do warn against inadequate use:

> 'Informative use of the hypertext is encouraged, but effects which are distracting or have no informative purpose (e.g., cute animations, flashing text) will not be used. The overriding principles being the accurate and effective presentation of the content and ease with which the information can be accessed.' (JMEM)

The fact that many journals are published both in html and pdf suggests that editors are often motivated by a desire to maintain the textuality of their journal as a printable publication. This is sometimes expressed in explicit terms:

> 'Papers are published by G3 in both PDF and HTML versions. The HTML version, which may include ancillary information such as animations, movies, virtual reality images, downloadable table of data, computer code, etc. that take advantage of the electronic medium, can be viewed using a standard World Wide Web browser. The PDF version may also be viewed through a web browser using the Adobe Acrobat pdf plug-in but may also be printed in color or black and white, producing a document comparable in quality to those published in paper journals.' (GGG)
>
> 'Papers should include both a printed and an electronic version.' (JHSR)
>
> 'The changes in this issue borrow three items from paper format that ... adapt well to the computer screen and enhance the ability of the reader to access information: black print on a white background, narrow columns, and leading.' (DERMOJ)

In summary, only just over a quarter of the e-journals in our sample seem to stress the digital nature of the journal and to encourage authors to take advantage of the possibilities offered by the digital format. Most e-journals start off very much in the same manner as a print journal, perhaps with an editorial stating the mission of the journal framed in terms of the scientific discipline, but without paying any attention whatsoever to the new format. The lack of emphasis given to the digital format indicates that it is often regarded by the editors as a more convenient and less expensive mode of distribution rather than as a new publishing medium that provides new opportunities for authors to represent and communicate their research findings.

4. Open access journals

The e-only journals studied in the previous section do not seem to have transformed the scientific article by benefiting from the properties of the digital format. Since these journals are from the the first stage of e-journals as described in section 1.5 on page 34, it might be argued that this explains our findings and that digitization has had more effect on later development stages. To what extent is this true? The second stage consists of existing printed journals that are only digitized (as exact copies of the printed version) to facilitate distribution and access. A quick check is sufficient to show that this is indeed the case and that these journals are not 'digital' in the sense that we use here.

What then, about the third stage, the new, e-only 'open access' journals that have developed in recent years? In chapter 2 we stated that the increasing success of open access publishing remains predicated on the 'traditional' genre of the scientific article and that open access journals usually follow the traditional article format, peer-review procedures etc. How true is this, and how do open access journals compare to the first stage of e-only scientific journals? In order to answer these questions, we carried out a survey of journals published by one of the most important open access publishers, BioMed Central (BMC).

From the 119 research journals published by BioMed Central in February 2005, we took a random sample of 24 titles (20%), listed in Table 5.21. For each journal we examined 10 research articles from 2004. If the 2004 volume did not contain 10 articles, we added journals first from the 2005 volume and then, if necessary to obtain 10 articles, from 2003.[35]

In this sample of 240 research articles, we found no examples of video or sound attachments. Most articles contained black & white or colour images as well as tables. Data attachments of some sort were found in only 26 articles (11%). Most of these contain questionnaires, data sequences or additional tables, usually in word, pdf or xls (spreadsheet) format. These attachments are described in more detail in table 5.22. It seems safe to conclude that in terms of content delivered by authors, the open access journals published by BioMed Central are mostly in line with the traditional article format as might be found

[35] If this procedure did not generate 10 articles, the journal was abandoned and the next journal in the alphabetical sequence was taken.

Table 5.21. Selected BMC journals

Annals of Clinical Microbiology and Antimicrobials	http://www.ann-clinmicrob.com/
BioMedical Engineering OnLine	http://www.biomedical-engineering-online.com/
BMC Biotechnology	http://www.biomedcentral.com/bmcbiotechnol/
BMC Clinical Pathology	http://www.biomedcentral.com/bmcclinpathol/
BMC Ecology	http://www.biomedcentral.com/bmcecol/
BMC Genetics	http://www.biomedcentral.com/bmcgenet/
BMC Medical Education	http://www.biomedcentral.com/bmcmededuc/
BMC Medical Research Methodology	http://www.biomedcentral.com/bmcmedresmethodol/
BMC Neurology	http://www.biomedcentral.com/bmcneurol/
BMC Oral Health	http://www.biomedcentral.com/bmcoralhealth/
BMC Pregnancy and Childbirth	http://www.biomedcentral.com/bmcpregnancychildbirth/
BMC Urology	http://www.biomedcentral.com/bmcurol/
Cell Communication and Signaling	http://www.biosignaling.com/
Critical Care	http://ccforum.com/
Filaria Journal	http://www.filariajournal.com/
Human Resources for Health	http://www.human-resources-health.com
International Journal of Health Geographics	http://www.ij-healthgeographics.com/
Journal of Immune Based Therapies and Vaccines	http://www.jibtherapies.com/
Journal of Neuroinflammation	http://www.jneuroinflammation.com/
Lipids in Health and Disease	http://www.lipidworld.com/
Molecular Cancer	http://www.molecular-cancer.com/
Reproductive Biology and Endocrinology	http://www.rbej.com/
Theoretical Biology and Medical Modelling	http://www.tbiomed.com/
World Journal of Surgical Oncology	http://www.wjso.com/

Table 5.22. Attachments in BMC journals

Type	Number of journals	Format
Data tables	12	pdf, word, xls
Questionnaires	7	word
Sequences and gene locations	4	pdf, word, xls
Literature search	2	word
Analysis charts	1	pdf

in any printed peer reviewed research journal. Even the rare instances where additional material is added are mostly non-substantial in format and volume, and might easily have been included in the body of the text.

Should one then conclude that the BMC journals present no innovative features in terms of utilizing the digital format? In order to answer this question it is not sufficient to limit the analysis to the article level. For if we look at the context provided at the aggregate level by the journal, and by BioMed Central as a publishing environment, it becomes clear that BMC has created a hybrid publishing environment that – on the basis of the author's submission – uses both a traditional print-publishing mode for the pdf versions and a more innovative mode for the html version.[36]

An overview of the various features of BioMed Central is given in figure 5.12. Most of these features apply specifically to the web-based version of the journals, but some features, such as the hyperlinked table of contents and hyperlinked references, are carried over to the pdf version (although they are, of course, lost in printing). The overview shows that BioMed Central is especially focused on reader-oriented features such as navigation, selecting and searching, current awareness, and (re)use of citations and references. These are features that are not provided by the journal as such, but by BMC as *publisher*, i.e. they are available to all journals within BMC's digital environment. Features at the level of the individual article do not require a different mode of writing by the author, but are based either on structural features added in the editorial process (e.g. hyperlinks, thumbnails) or on BMC-wide functions (searching, citation/reference downloading, comments).

Although other studies show that users tend to appreciate enhancements of the type offered by BioMed Central,[37] it is difficult to assess the use that is made of them at any level of specificity. There is one case that we were able to check: the extent to which readers use the possibility to post comments on the articles that are published in BMC journals. Posting comments is relatively easy to do, although it does require a one-time free registration to the journal or BMC. However, in all 240 articles we found only one instance of users (2) adding comments to an article,[38] and one other comment (addendum) by the author of the original paper.[39]

To conclude this section, we observe that our survey of open access journals shows that they have in no way transformed the research article by incorporating specific digital properties. What we do see is that at the aggregate level

[36] Most journals published by BioMed Central are available in electronic form only, but some journals also have a print edition. BioMed Central makes *printed* archival editions of all other journals available on demand at the end of each calendar year.

[37] Rusch-Feja and Siebeky 1999; Voorbij 2005.

[38] Two marginal comments on the use of red and green, colours that are difficult to distinguish by readers suffering from so-called 'red-green color blindness' (http://www.ij-healthgeographics.com/content/3/1/10/comments).

[39] http://www.tbiomed.com/content/1/1/10/comments.

Article level:

- Hyperlinked table of contents
- Hyperlinks to/from references
- References hyperlinked to PubMed and publisher (abstract or full text when available)
- Hyperlinks to figures, tables (in separate window)
- Figure/table thumbnails in margin linked to figure in low and high resolution (in separate window)
- Search PubMed for related articles
- Search PubMed for authors of article
- Email article to friend
- Download citation/references (various formats)
- Post comments to article

Aggregate level:

- Advanced search options
- Select article type (e.g. research, case report, editorial, review, debate, technical innovations)
- Select 10 top most accessed articles
- My BioMed Central / My [journal] (stored searches, recent/highlighted articles, topical selections)
- Find related journals
- Jump to volume/art.nr.
- RSS feed of latest articles
- Download papers to pda (via AvantGo)

Figure 5.12. Features of BMC journals

BMC has developed a number of features aimed at offering the *user* a more functional environment for accessing and reading the scientific literature.

Anyone is free:

* to copy, distribute, and display the work;
* to make derivative works;
* to make commercial use of the work;

Under the following conditions: Attribution

* the original author must be given credit;
* for any reuse or distribution, it must be made clear to others what the license terms of this work are;
* any of these conditions can be waived if the authors gives permission.

Statutory fair use and other rights are in no way affected by the above.

(source: http://www.biomedcentral.com/info/about/openaccess/)

Figure 5.13. BMC's Open Access policy

5. Evaluation

The results of our survey can be summarized as follows:

Formats Submission formats are predominantly based on word processing; authors are seldom required to use more web-oriented formats such as html. Over 2/3 of the journals is published in html, but often in combination with other formats, especially pdf. In most cases the author's submission is transformed to a web format through the editorial process, but clearly less so in the sciences (where "camera" ready submission is more common) than in the humanities.

Multimedia Colour images, graphs etc. were use in 25% of the journals. Other types of multimedia such as audio, video, software and animations were found in only 16%. Even in journals where we did find multimedia, their use was almost invariably infrequent, often in spite of the fact that authors were encouraged to include multimedia by the journal's editors.

Data resources The use of data resources included in or linked to journal articles is very rare: only 9% of the journals contain examples, and even there they occur infrequently. Most examples are from the sciences.

Revision In general, e-journals follow the prevailing practice that articles, once published, are final and are not open to revision by the author.

Response With very few exceptions, e-journals do not succeed in involving the reader in an interactive dialogue or even in soliciting comments that could be attached to the original article.

Customization E-journals do not use characteristics of the reader to adapt content or presentation. At the aggregate level, some customization is offered in the form of "personal" folders that allow the user to store selected articles for subsequent use. Another form of personalization is the alerting service that sends an e-mail to subscribers when new articles appear in the journal.

External hyperlinks The most frequently used publication formats (html and pdf) allow external hyperlinks (urls), but in a large number of cases (40%) these are not used or not (consistently) activated. There were no examples of typed linking.

Functionality There are almost no examples to be found "functional" documents (i.e. articles with embedded functionality). There are, however, various examples of user-oriented functions at the aggregate level of the journal. The main functionality found in our sample is the search function.

Navigation E-journals contain various examples of devices that assist the reader in navigating the journal and its articles. We found 105 cases (56%) where journals use such devices. However, the majority of the journals use relatively simple means such as hyperlinks to/from notes or references, or hyperlinked content tables. Of the journals published in html format, 28% do not contain internal navigational hyperlinks. When they do, these links are often not retained in parallel pdf versions.

Table 5.23 provides an overview of the editorial policies found in our survey.of our survey.

The general impression that one gets from these results is that the e-only journals published in the period 1987-2004 have not used the specific digital features of the e-journal to a large extent. Apart from benefits such as low cost and rapid dissemination, most e-journals are not significantly different from print journals with respect to specific properties such as presentation, interactivity and functionality. In terms of our research model (fig. 5.1), the possibilities inherent in the digital format are not fully realized. This can be ascribed to two distinct factors:

1. A lack of utilization of digital features by authors (e.g. reluctance to include multimedia content, add data resources, engage in interactive discourse, etc.), even when these features are available and encouraged by the journal.

2. Rules and prescriptions, imposed by journals and their editors, that limit the possibilities available to authors. Here, two sub-factors are at play:

    - a lack of implementation of advanced features (e.g. novel peer review methods, mechanisms for revision and response, inclusion/activation of hyperlinks, navigational devices, and advanced functionality)
    - restrictive editorial policies (e.g. limits on length, content elements etc.; enforcement of print-based style rules and typography) and general lack of recognition of the digital nature of the journal and of encouragement of authors to utilize the new possibilities.

Table 5.23. Summary of editorial policies

Style rules	Most journals enforce style rules, either proprietary or derived from existing style manuals such as APA and MLA. Style instructions are often very detailed and tailored toward print-style publishing. In general, style and structure seem to be governed by the traditional rules prevalent in the scientific discipline.
Limits	Although some journals recognize the size restrictions are not economically necessary for digital publishing, many do impose size limits based on the belief that brevity is appreciated by readers of online publications. There are also many limits on the use of footnotes, figures and other elements.
Typography	Many journals discourage the use of explicit formatting by authors unless camera-ready copy (postscript/pdf) is required. Formatting (and the transformation of non-webbased forms to html) is performed through the editorial process.
Digitality	Explicit mentioning of the digital nature of the journal and consequences/guidelines for authors was found in only 50 journals (27%). These editorial statements tend to be framed in positive terms, emphasizing the added value and new possibilities, especially in terms of multimedia representation, hyperlinks and inclusion of large data quantities.

We conclude this chapter with an evaluation of our survey results in terms of the digital characteristics specified in table 5.3. The evaluation is summarized in table 5.24.

Multimedia content

Less than a quarter (22.5%) of the e-journals contain multimedia content other than (B/W or colour) images. Even in these journals, articles that actually contain multimedia content are rare. We conclude therefore that the multimedia capabilities of the digital format are not regarded as essential by most e-journals, and that authors are not inclined to include multimedia content in their scientific writings. The textual form of presentation is maintained in the digital environment.

Table 5.24. Summary of conclusions

Digital characteristics	Article properties
Multimedia content	Most e-journals do not regard multimedia capabilities as essential; authors are not inclined to including multimedia content.
Network access	All journals studied were available over the Internet, mostly without access restrictions.
Network connectivity	Network connectivity in scientific e-journals is relatively weak, and not universally regarded as an essential property of the e-journal.
Author control	In general, e-journals follow the prevailing practice that articles, once published, are final and are not open to revision by the author.
Dynamic content	The content of e-journals is predominantly static.
Adaptability	Adaptation strategies as such are virtually absent from e-journals.
Functionality	Embedded functionality is almost entirely lacking except for basic search and navigation mechanisms.
Copyability	Specific mechanisms that would limit copyability are not used.
Reader control	Mechanisms for creating a more interactive reading experience are rarely used. In general, the reader has little or no control over content and presentation of the journal.
Flexibility	Some flexibility in areas such as length, number of graphical elements allowed, etc. In general the e-journal has not resulted in a significantly grater flexibility in terms of periodicity, size, typography etc.

Network access

All journals studied were available via the Internet. In almost all cases access was not restricted. Less than 5% was subscription based requiring a password or institutional subscription for full access.

Network connectivity

Network connectivity is related to the the use of external hyperlinks to resources such as cited literature, background information, data resources etc. Although the most frequently used publication formats (html and pdf) allow external hyperlinks (url's), these are not used or not (consistently) activated in a large number of cases (40%). Also, the use of data resources included in or linked to journal articles is very rare: only 9% of the journals contain examples, and even there they occur infrequently. Pdf-versions (often in parallel with html) are often produced without activated hyperlinks. We conclude therefore that network connectivity in scientific e-journals is relatively weak, and not universally regarded as an essential property of the e-journal.

Author control

In the majority of cases the author's submission is transformed to its final form through some kind of editorial process over which the author has no control. As we have seen, authors are sometimes even discouraged to add any kind of formatting to their submissions. The extent to which the author has control over the published article is related to a number of factors. Direct revision is usually impossible since, although journal articles are usually stored at a well-defined and single location (as far as we could establish, mirroring is very rare), authors do not obtain write access to journal articles. Another possibility would be a mechanism offered by the journal to revise the text after publication, or at least to link the text to subsequent corrections and revisions. As with print journals, if necessary authors can provide corrections in a subsequent issue of the e-journal. Backlinking of corrections to the original article is sometimes added. However, we found only very few examples where authors were allowed to maintain different versions. In general, therefore, e-journals follow the prevailing practice that articles, once published, are final and are not open to revision by the author. This is also illustrated by the mostly traditional approach of e-journals to copyright, where rights are normally either implicitly or explicitly held by the journal. Even in cases where authors retain copyright, they are expected to refer to 'the original publication' in subsequent publications, reflecting the traditional idea of a publication that is 'final'.

Dynamic content

The lack of author control is one aspect that illustrates the more general lack of dynamic properties in e-journals and the prevalent concept of 'final' publication. We found no convincing examples of articles that contain dynamic content to monitor change or adapt to new information. The content of e-journals is, therefore, predominantly static. The dynamic property that we have identified in chapter 4 as being an important characteristic of the digital format is strikingly absent from scientific e-journals.

Adaptability

We have defined adaptability as the availability of strategies to adapt form, content and/or functionality of the article to the context, e.g. characteristics of the user. We found no examples where a journal actively solicits user characteristics to which it can adapt itself. We did find a few cases where the user can actively change various presentation characteristics, but we list these under 'reader control' below. In fact, we found only one case where the user could set parameters that will change the visual presentation of the article itself[40] (rather than applying changes to the rendering mechanism, e.g. the browser). Adaptation strategies as such are, therefore, virtually absent from e-journals.

Functionality

We found no examples of semantic linking or adaptive hypermedia, and only very few examples of embedded software, database interfacing, etc. There are, however, various examples of user-oriented functions at the aggregate level of the journal. The main functionality found in our sample is the search function. There are also examples of navigational devices, but these are generally very basic (e.g. hyperlinked tables of contents). Advanced embedded functionality is almost entirely lacking. Again, a property that we have identified in chapter 4 as being an important characteristic of the digital format is almost completely absent from scientific e-journals.

Copyability

Although many e-journals are protected by copyright, we found no specific mechanisms (e.g. Digital Rights Management) that would limit copyability, or other examples where it was not possible to download articles.

Reader control

As mentioned above, we did find cases where the user can actively change various presentation characteristics, although these cases are rare and are usually

[40]Internet Journal of Chemistry, see table 5.15.

limited to a small set of choices (e.g. framed/frameless presentation,) In addition, many journals offer a choice between html and pdf. Various devices are sometimes available that allow the reader to navigate his/her own path through the article (e.g. hyperlinked tables of contents, navigation panels). Also, the possibility to create a more interactive reading experience (e.g. by responding to an article and/or interact with the authors or editors), is usually not available. In general, the reader has little or no control over content and presentation of the journal, and hyperlinked navigational devices that are almost ubiquitous on the WWW are often absent.

Flexibility

The vast majority of e-journals follow a traditional periodical publication schema (e.g. volumes and issues) rather than adding articles on a ongoing basis. Novel approaches to peer review are virtually absent. Copyright policies are mostly traditional, disallowing prior or parallel publication, although a number of journals do allow the author to retain copyright if the first publication is acknowledged in subsequent publications. Editorial policies are often surprisingly strict, prescribing exact details such as margins and font. Within a journal a uniform typography is usually enforced. To a certain extent there is greater flexibility than with print in areas such as length, number of graphical elements allowed, etc. But in general the e-journal has not resulted in a significantly greater flexibility in terms of organization, periodicity, size, typography etc.

6. The impact of digitization on the scientific article

In 1999 the American Quarterly launched an experiment in hypertext publishing,[41] together with the American Studies Crossroads Project[42] (Georgetown University) and the Center for History & New Media at George Mason University[43]. The four articles contained in the experiment, heavily hypertexted and laden with film fragments, sound files, and even a virtual bedroom, formed a radical break away from the traditional text-oriented print style of the journal.[44] On-line publication in hypertext format was apparently too startling an experience for AQ: the experiment has not been repeated. The story of AQ's project can serve as an example of scientific communication in general: there have been a number of experiments in digitization aimed at creating new forms of knowledge representation and communication. But to a large extent

[41] See http://chnm.gmu.edu/aq/ and http://muse.jhu.edu/journals/american_quarterly/toc/aq51.2.html. The call for proposals can be found at http://www.georgetown.edu/crossroads/expo/aquarterly.html.

[42] http://www.georgetown.edu/crossroads/.

[43] http://chnm.gmu.edu/.

[44] Poster 2001, p. 95.

the peer-reviewed scientific article has remained unaffected by these experiments and continues, as we have seen, very much as it used to be.

The results of the study described in this chapter can therefore serve as a useful reality-check for various expectations that have been voiced as to the role of digitization for the development of the scientific journal. Many authors describe this role, as we have seen, in terms of a 'digital revolution' in scientific communication – often without specifying the precise meaning of what might be expected from such an event. Other authors, however, are more specific and discuss particular features that they expect from digitization of the research journal. For instance, Lancaster (1995) described the advantages of electronic publishing as rapid and efficient dissemination, innovative presentation of research results, public peer review, and low publishing cost. Nentwich (2003), in his study on 'cyberscience', lists a number of advantages of e-publishing that are closely related to the variables used in our own study (table 5.25).[45] In another extensive survey of the literature, Friedlander and Bessette (2003, p. 22) identified an array of specific features that were expected to appear in the e-journal (table 5.26). McKiernan (2002, p. 296 ff) contends that 'an ever increasing number of electronic journals are transcending the limitations of the paper medium by incorporating and integrating a wide variety of innovative electronic features and content', and gives a list of these features under a number of interesting headings (table 5.27). To give a final example, Tenopir and King (2000, p. 349) express their positive expectations about the digital environment for scientific publishing (including both journals and other communication forums) in terms of specific characteristics:

> 'We contend that in the context of scientific scholarly publishing, the Web is much more than a distribution medium because it incorporates two key electronic elements which have the potential to revolutionize the scholarly communication system: (1) the use of multimedia applications, and (2) interactivity between authors and readers'.

Although we have seen incidental examples of most of these features, they are not characteristic of the majority of articles published in e-only journals, let alone of the far greater number of scientific journals that are published as digital copies of print journals.

This study is not the first to address the impact of digitization on the scientific journal empirically rather than on a speculative basis, though it is probably the most thorough one. A relatively early – and rather superficial – survey by Singh et al. (1998) states that:

> 'We could not trace any Interactive Multi-Media Journals. ... There are very few electronic-only or even electronic versions of print-based engineering journals. It appears that most of the electronic journals got a boost primarily out of a desire to reduce the time between submission and publication. ... most of them are essentially a static visual form of their counterpart hard-copy

[45]Nentwich (2003, section 7.2).

Table 5.25. Advantages of e-publishing

■ Improved navigation in articles
■ Higher speed of publishing and dissemination
■ Enhanced forms of layout
■ New forms of multimedia content
■ Linking between articles and sources
■ Greater flexibility in length of articles
■ Search function
■ Possibilities for hypertext
■ Transforms fixity
■ Adds interactivity
■ Leads to new types of publications
■ Allows for new forms of refereeing
(Nentwich 2003, ch 7, p. 323)

journal. There are 'electronic only journals' too, but with hardly any mentionable superiority in terms of appearance and substance.'

Burg, Wong, Yip, and Boyle (2000), editors of the Interactive Multimedia Electronic Journal of Computer-Enhanced Learning, carried out an informal review of the state-of-the-art in interactive multimedia journals for academia and draw a similar conclusion:

> 'For over a decade, academics have been predicting a revolution in scholarly publication. ... Nevertheless, a closer look at existing online journals shows that they fall short of the predictions.
>
> We expected to find a good amount of multimedia material in scientific and medical publications, where the usefulness of 3-D images and simulations is easy to imagine, but our findings were disappointing. ... Overall, it is difficult to find examples of the type of journal predicted since the early 1990's. Even the Journal of Interactive Media in Education (http://www-jime.open.ac.uk), an early and well-produced entry into the field of multimedia publication, appears to be backing away from multimedia interactivity in recent issues.'

It is not only the case that publishers and editors of e-journals are reluctant to exploit the possibilities of the digital format for their content: authors and users are also reluctant to utilize those possibilities. We found in our study that many features that have been announced and made available by the digital journal are not used by its authors and readers, or only very infrequently. For instance, as

Table 5.26. Expected features of e-journal articles

- animation and virtual reality
- use of color
- support for mathematical and chemical notation
- hyperlinks to other articles, supporting evidence, algorithms
- visualization
- multimedia and interactive displays
- incremental publication (i.e. as soon as the publication is ready)
- user-defined collections based on user preferences
- updating, flexibility, adaptable to change
- access to search mechanisms
- facilities for enhanced interactivity and discourse among readers and authors
- improved access
- improved quality ranking

(from: Friedlander and Bessette 2003, p. 22, summarizing a large number of other publications)

Table 5.27. E-journal features according to McKiernan

Edition	electronic submission, refereeing and review; virtual e-journals; synoptic e-journals
Ego-centric	alerting services; personalized e-journals; font, format and display control
Electric	indexing and searching; computer code; translation services
Empowering	download options; reader participation; virtual filing cabinets (managing relevant citations with links to abstracts and full-text)
Entwined	reference linking; citation indexing; relatedness (similarity searching)
Explorative	database linking (i.e. from database citation to full text publication); demonstrations/multimedia: animation, streaming and non-streaming audio and video, 3D-interactive models
Expressive	discussion forums; dynamic articles (augmenting a previously published work with current findings and new observations); reactive e-journals (user response)
Extra	access to restricted databases; e-book access; supplemental data

we have seen, although many journals allow the inclusion of various types of multimedia, these are often not used in the articles published.

The non-use of available digital features can be subdivided into two categories. One is where authors are asked to provide materials that normally would not be included in a printed publication, e.g. various types of multimedia. The other is where actors in the information chain (e.g. authors, reviewers and readers) are asked to engage in a level of (inter)action that again is specific to the digital format: updating (authors), discussion (reviewers) and responding to articles (readers). In all we found 22 journals in our sample that promised features that did not materialize in the publication.[46] In most other cases, the use of these features was usually the exception rather than the rule.

We can conclude, therefore, that the e-journal does not fully benefit from the possibilities of the digital format, and that even in the minority of cases where the journal specifically presents itself as an innovative, digital medium and the editors encourage authors to utilize its specific properties, the advertised features are often not utilized at all or only very infrequently. As a result, the specific properties of the digital format that we have described in chapter four are reflected only to a very limited degree in the contents of the scientific e-journal. More innovative features are to be found at the aggregate level where publishers and other content suppliers add functions for searching, browsing and linking over larger collections of journals and articles. However, these features barely impact on the form and substance of the article itself when it is published in a digital journal.

What we see, then, is that our study reveals a large difference between expectations and reality or – in terms of our research model – possibilities and realization. We have to conclude that the impact of digitization on the scientific article is as yet very limited. What authors send into the world, and the content that is made available through the electronic journal, remains very much the same as what traditionally has been published in print. One might perhaps expect that at least those authors who actively choose to publish in an electronic journal might be more inclined to adopt the specific digital properties of the journal in their writings. But this is not the case.

In view of this finding and of the fact that the volume of e-only journals is also extremely low in comparison with the traditional print journal (and their digital versions), we conclude that the e-journal has not transformed formal scientific communication, and that claims as to 'revolutionary' effects of e-publishing cannot be substantiated. Indeed, as Uren et al. (2003) notes: 'the WWW has not yet impacted on one of the primary activities in research:

[46]Hyperlinks are not included in this count since their use varies widely between articles. As we have seen, hyperlinks are often not consistently activated even in html-based journals.

assessing new findings in the light of current knowledge and debating it with colleagues'.

Nevertheless, it is unquestionable that the scientific journal has become 'digital' and that the 'e-journal' is quickly replacing the traditional print journal. A large and increasing proportion of journals are available in digital form and can be accessed over the network through publishers, libraries or aggregators. Many if not most scientists - at least in the Western world and especially in the sciences and social sciences - are now beginning to regard the scientific journal as a digital publication format. We shall therefore have to explain the seemingly paradoxical situation that digitization has had little impact on the scientific *article*, whereas the digital *journal* is becoming the most important format for formal scientific communication.

Appendix: Digital journals list

code	title	url
ACL	Academic leadership	http://www.academicleadership.org/
ACMJEA	ACM journal of experimental algorithmics	http://www.cs.amherst.edu/newjea/
AHR	Australian humanities review	http://www.lib.latrobe.edu.au/AHR/
AJDTS	Australasian journal of disaster and trauma studies	http://www.massey.ac.nz/~trauma/
AMEN	Applied mathematics e-notes	http://www.math.nthu.edu.tw/~amen/
ANCNAR	Ancient narrative	http://www.ancientnarrative.com/
ANTHR	Anthropoetics	http://www.anthropoetics.ucla.edu/
AOIJ	Academic open internet journal	http://www.acadjournal.com
APPLES	Applied language studies	http://www.solki.jyu.fi/apples/
APPS	Applied semiotics	http://www.chass.utoronto.ca/french/as-sa/
APS	AAPS PharmSci	http://www.aapspharmsci.org/
APST	AAPS PharmSciTech	http://www.aapspharmscitech.org/
ARLO	Acoustics research letters online ARLO	http://ojps.aip.org/ARLO/
ARSDIS	Ars disputandi	http://www.arsdisputandi.org/
ASQ	African studies quarterly	http://www.africa.ufl.edu/asq/

code	title	url
ATMOSL	Atmospheric science letters	http://www.sciencedirect.com/science/journal/1530261X
BEJE	Brazilian electronic journal of economics	http://www.beje.decon.ufpe.br/
BPO	Biological procedures online	http://www.biologicalprocedures.com/bpo/general/home.htm
BQEST	B¿Quest	http://www.westga.edu/~bquest/year.html
CECOMM	CrystEngComm	http://www.rsc.org/is/journals/current/crystengcomm/cecpub.htm
CGD	Conformal geometry and dynamics - an electronic journal of the AMS	http://www.ams.org/ecgd/
CHEMED	Chemical educator	http://www.chemeducator.org/
CHEMJI	Chemical journal on internet	http://www.chemistrymag.org/
CIE	Current issues in education	http://cie.ed.asu.edu/
CJEAP	Canadian journal of educational administration and policy	http://www.umanitoba.ca/publications/cjeap/
CLCWEB	Comparative literature and culture	http://clcwebjournal.lib.purdue.edu/
CONZO	Contributions to zoology	http://clcwebjournal.lib.purdue.edu/
CRISP	Current research in social psychology	http://www.uiowa.edu/~grpproc/crisp/crisp.html
CROMOHS	CROMOHS: cyber review of modern historiography	http://www.cromohs.unifi.it
CTHEOR	CTheory	http://www.ctheory.net/
CULTL	Cultural logic	http://eserver.org/clogic/
CULTM	Culture machine	http://culturemachine.tees.ac.uk/frm_f1.htm
CURREL	Currents in Eelectronic literacy	http://www.cwrl.utexas.edu/currents/
CYBM	Cybermetrics	http://www.cindoc.csic.es/cybermetrics/
DERMOJ	Dermatology online journal	http://dermatology.cdlib.org/
DGDS	Differential geometry - dynamical systems	http://vectron.mathem.pub.ro/dgds/
DJOPHT	Digital journal of ophthalmology	http://www.djo.harvard.edu/

code	title	url
DMTCS	Discrete mathematics and theoretical computer science	http://dmtcs.loria.fr/
DOCMATH	Documenta mathematica	http://www.mathematik.uni-bielefeld.de/documenta/Welcome-eng.html
DTLJ	Digital technology law journal	http://wwwlaw.murdoch.edu.au/dtlj/
EAR	Early American review	http://www.earlyamerica.com/review/
EARTHINT	Earth interactions	http://earthinteractions.org/
ECHO	Echo - a music-centered journal	http://www.echo.ucla.edu/
ECM	European cells and materials	http://www.ecmjournal.org/
ECOSOC	Ecology and society	http: //www.ecologyandsociety.org/
ECRP	Early childhood research & practice	http://ecrp.uiuc.edu/
EDTS	Educational technology & society	http://ifets.ieee.org/periodical/
EGJ	Electronic green journal	http: //egj.lib.uidaho.edu/index.html
EIOP	European integration online papers	http://eiop.or.at/eiop/
EJANZ	Electronic journal of Australian and New Zealand history	http://www.jcu.edu.au/aff/history/index.htm
EJAP	Electronic journal of analytic philosophy	http://ejap.louisiana.edu/archives.html
EJBIOT	Electronic journal of biotechnology	http: //www.ejbiotechnology.info/
EJC	Electronic journal of communication	http://www.cios.org/www/ejcmain.htm
EJCBS	Electronic journal of cognitive and brain sciences	http: //www.ejcbs.com/ejcbs.html
EJCJS	Electronic journal of contemporary japanese studies	http: //www.japanesestudies.org.uk/
EJCL	Electronic journal of comparative law	http://law.kub.nl/ejcl/
EJCOMB	Electronic journal of combinatorics	http://www.combinatorics.org/
EJCP	Electronic journal of communicative psychoanalysis	http://www.ejcpsa.com/
EJDE	Electronic journal of differential equations	http://ejde.math.swt.edu/

code	title	url
EJGE	Electronic journal of geotechnical engineering	http://www.ejge.com/ index_ejge.htm
EJHS	Electronic journal of human sexuality	http://www.ejhs.org/
EJIST	E-journal of instructional science and technology	http://www.usq.edu.au/ electpub/e-jist/
EJITC	Electronic journal of information technology in construction	http://www.itcon.org/
EJLA	Electronic journal of linear algebra	http://www.math.technion.ac. il/iic/ela/
EJO	Electronic journal of oncology	http://www.elecjoncol.org/
EJOS	Electronic journal of oriental studies	http://www2.let.uu.nl/Solis/ anpt/ejos/EJOS-1.html
EJOUR	Ejournal	http: //www.ucalgary.ca/ejournal/
EJPECP	Electronic journal of probability	http://www.math.washington. edu/~ejpecp/
EJPH	Electronic journal of pathology and histology	http: //ejpath.amu.edu.pl/index.html
EJQTDE	Electronic journal of qualitative theory of differential equations	http://www.math.u-szeged.hu/ ejqtde/
EJROT	Electronic journal of radical organisation theory	http://www.mngt.waikato.ac. nz/ejrot/
ELANT	Electronic antiquity	http://scholar.lib.vt.edu/ ejournals/ElAnt/
ELAW	E law	http: //www.murdoch.edu.au/elaw/
EMLS	Early modern literary studies	http://www.shu.ac.uk/emls/ emlshome.html
ENTROP	Entropy	http://www.mdpi.net/entropy/
EOL	Ethnomusicology online	http://research.umbc.edu/ efhm/eol.html
EPAA	Educational policy analysis archives	http://epaa.asu.edu/epaa
ETNA	Electronic transactions on numerical analysis	http://etna.mcs.kent.edu/
EUNOMIOS	EUNOMIOS: an open online journal for theory, analysis and semiotics of music	http://www.eunomios.org/
FM	First monday	http://www.firstmonday.dk/

code	title	url
FQS	FQS Forum qualitative research	http://qualitative-research.net/fqs/fqs-eng.htm
GEOCHEM	Geochemical transactions	http://gt.aip.org/gt/?jsessionid=706841086362079823
GEOTOP	Geometry and topology	http://www.maths.warwick.ac.uk/gt/
GGG	Geochemistry, Geophysics, Geosystems	http: //www.agu.org/journals/gc/
GIDA	Journal of geographic information and decision analysis	http://www.geodec.org/
HAPTICS	Haptics-e: the electronic journal of haptics research	http://www.haptics-e.org/
HYLE	International journal for philosophy of chemistry	http://www.hyle.org/
IEJHE	International electronic journal of health education	http://www.aahperd.org/iejhe/
IEJLL	International electronic journal for leadership in learning	http://www.ucalgary.ca/~iejll/
IJAM	Internet Journal of Airway Management	http://www.ijam.at/
IJANES	Internet journal of anesthesiology	http://www.ispub.com/ostia/index.php?xmlFilePath=journals/ija/front.xml
IJC	Internet journal of chemistry	http://www.ijc.com/IJC/
IJEICM	Internet journal of emergency and intensive care medicine	http://www.ispub.com/ostia/index.php?xmlFilePath=journals/ijeicm/current.xml
IJFD	International journal of fluid dynamics	http: //elecpress.monash.edu.au/ijfd/
IJIC	International journal of integrated care	http://www.ijic.org/
IJMS	International journal of molecular sciences	http://www.mdpi.net/ijms/
IJMT	Internet journal of medical toxicology	http://www.ijmt.net/
IJRDM	Internet Journal of Rescue and Disaster Medicine	http://www.ispub.com/ostia/index.php?xmlFilePath=journals/ijrdm/current.xml
IJTCS	Internet journal of thoracic and cardiovascular surgery	http://www.ispub.com/ostia/index.php?xmlFilePath=journals/ijtcvs/front.xml

code	title	url
IMEJ	Interactive multimedia electronic journal of computer-enhanced learning	http://imej.wfu.edu/
INFRES	Information research	http: //informationr.net/ir/index.html
INTARCH	Internet archaeology	http://intarch.ac.uk
INTERSEC	Intersections: gender, history and culture in the Asian context	http://wwwsshe.murdoch.edu.au/intersections/
INTERSTAT	Interstat	http://interstat.stat.vt.edu/InterStat/intro.html-ssi
ISTL	Issues in science and technology libraries	http://www.istl.org/
JAHC	Journal of the Association for History and Computing	http://mcel.pacificu.edu/JAHC/jahcindex.htm
JAIR	Journal of artificial intelligence research	http://www.cs.washington.edu/research/jair/
JAIS	Journal of Arabic and Islamic studies	http://www.uib.no/jais/
JASSS	Journal of artificial societies and social simulation	http://jasss.soc.surrey.ac.uk/JASSS.html
JCJPC	Journal of criminal justice and popular culture	http://www.albany.edu/scj/jcjpc/index.html
JCMC	Journal of computer-mediated communication	http://www.ascusc.org/jcmc/
JCS	Journal of cotton science	http://journal.cotton.org/
JCSE	Journal of corrosion science and engineering	http://www2.umist.ac.uk/corrosion/JCSE/
JCTR	Journal for christian theological research	http://home.apu.edu/~CTRF/jctr.html
JDC	Journal of design communication	http://scholar.lib.vt.edu/ejournals/JDC/
JEGMS	Journal of empirical generalisations in marketing science	http://www.empgens.com/
JEMIE	Journal on ethnopolitics and minority issues in europe	http://www.ecmi.de/jemie/
JFLP	Journal of functional and logic programming	http://scholar.lib.vt.edu/ejournals/JFLP
JHEP	Journal of high energy physics	http://jhep.sissa.it/
JHS	Journal of Hebrew scriptures	http: //www.arts.ualberta.ca/JHS

code	title	url
JHSR	Journal of hazardous substance research	http://www.engg.ksu.edu/HSRC/JHSR
JILT	Journal of information, law and technology	http://elj.warwick.ac.uk/Jilt/
JIME	Journal of interactive media in education	http://www-jime.open.ac.uk/
JIPAM	Journal of inequalities in pure and applied mathematics	http://jipam.vu.edu.au/
JIS	Cal Poly pomona journal of interdisciplinary studies	http: //www.csupomona.edu/~jis/
JIS	Journal of integer sequences	http://www.cs.uwaterloo.ca/journals/JIS/
JITTA	Journal of information technology theory and application	http://64.247.32.28/
JMEM	Journal of memetics	http://jom-emit.cfpm.org/
JMR	Journal for maritime research	http://www.jmr.nmm.ac.uk/
JNCN	Journal of neurology and clinical neurophysiology	http://www.ncnpjournal.com/
JODI	Journal of digital information	http://jodi.ecs.soton.ac.uk/
JOE	Journal of extension	http://www.joe.org/
JOP	Journal of the pancreas	http://www.joplink.net/
JOUVERT	Jouvert: a journal of postcolonial studies	http://152.1.96.5/jouvert/
JPPS	Journal of pharmacy & pharmaceutical sciences	http://www.ualberta.ca/~csps/Journals/JPPS.htm
JRS	Journal of religion and society	http: //moses.creighton.edu/JRS/
JSCM	Journal of seventeenth-century music	http://sscm-jscm.press.uiuc.edu/jscm/
JSE	Journal of statistics education	http://www.amstat.org/publications/jse/
JUCS	Journal of universal computer science	http://www.jucs.org/
JVE	Journal of virtual environments	http: //www.brandeis.edu/pubs/jove/
KACIKE	Kacike: journal of Caribbean Amerindian history and anthropology	http://www.kacike.org/
LGD	Law, social justice and global development	http: //elj.warwick.ac.uk/Global/

code	title	url
LIBRES	Library and information science research	http://libres.curtin.edu.au/
LINGON	Linguistik online	http: //www.linguistik-online.de
LMS	LMS journal of computation and mathematics	http://www.lms.ac.uk/jcm/
LORE	Lore	http://www-rohan.sdsu.edu/ dept/drwswebb/lore/lore.html
LPP	Library philosophy and practice	http://www.webpages.uidaho. edu/~mbolin/lpp.htm
LRR	Living reviews in relativity	http: //relativity.livingreviews.org/
LRR	Living reviews in relativity	http: //relativity.livingreviews.org/
MA	Music and anthropology	http://research.umbc.edu/eol/ MA/index.htm
MEDED	Medical education online	http: //www.med-ed-online.org/
MOLEC	Molecules	http: //www.mdpi.net/molecules
MOLVIS	Molecular vision	http://www.molvis.org/molvis/
MPEJ	Mathematical physics electronic journal	http://www.ma.utexas.edu/ mpej/MPEJ.html
MRS	MRS internet journal of nitride semiconductor research	http://nsr.mij.mrs.org/
MTO	Music theory online	http://www. societymusictheory.org/mto/
NEXUS	Nexus network journal	http: //www.emis.de/journals/NNJ/
NHAE	New horizons in adult education	http://www.nova.edu/~aed/ newhorizons.html
NJC	New journal of chemistry	http://www.rsc.org/is/journals/ current/newjchem/njc.htm
NJP	New journal of physics	http://www.iop.org/EJ/journal/ 1367-2630/1
NYJM	New York journal of mathematics	http://nyjm.albany.edu: 8000/nyjm.html
OE	Optics express: the international electronic journal of optics	http://www.opticsexpress.org/
OJC	Online journal of cardiology	http://sprojects.mmi.mcgill.ca/ heart/

code	title	url
OJIN	Online journal of issues in nursing	http://www.nursingworld.org/ojin/index.htm
OJVR	Online journal of veterinary research	http://www.cpb.ouhsc.edu/OJVR/jvet196a.htm
OPTIC	Optics express	http://www.opticsexpress.org/
PE	Palaeontologia electronica	http://palaeo-electronica.org/
PHIN	Philologie im Netz	http://www.fu-berlin.de/phin/
PMC	Postmodern culture	http://muse.jhu.edu/journals/postmodern_culture/
PSYCHE	Psyche	http: //psyche.cs.monash.edu.au/
PSYCHO	Psycholoquy	http://www.cogsci.soton.ac.uk/psycoloquy/
QALREP	Qualitative report	http://www.nova.edu/ssss/QR/index.html
RADPED	Radical pedagogy	http: //radicalpedagogy.icaap.org
RENFOR	Renaissance forum	http://www.hull.ac.uk/Hull/EL_Web/renforum/
REPTH	Representation theory - an electronic journal of the AMS	http://www.ams.org/ert/
ROMNET	Romanticism on the net	http://www.ron.umontreal.ca/
SCOPE	SCOPE	http://www.nottingham.ac.uk/film/journal/
SINC	Sincronia: an e-journal of cultural studies	http://fuentes.csh.udg.mx/CUCSH/Sincronia/index.html
SRO	Sociological research online	http: //www.socresonline.org.uk/
STLR	Stanford technology law review	http://stlr.stanford.edu/STLR/Core_Page/index.htm
SWJPAM	Southwest journal of pure and applied mathematics	http://rattler.cameron.edu/swjpam/swjpam.html
TAC	Theory and applications of categories	http://www.tac.mta.ca/tac/
TESL	TESL-EJ	http://www-writing.berkeley.edu/TESL-EJ
TRANS	TRANS	http://www.inst.at/trans/
WCR	Western criminology review	http://wcr.sonoma.edu/
WJCLI	Web journal of current legal issues	http://webjcli.ncl.ac.uk/

Chapter 6

DIGITIZATION AND THE EVOLUTION OF SCIENTIFIC COMMUNICATION

In this study we have sketched the development of scientific communication and the scientific journal over the centuries, up to the e-journals of our times (chapter 2), and analyzed in depth the use made of digital means by the authors of peer-reviewed articles. To go beyond mere description, we must now address the question *why* scientific communication has developed in the way it has. More specifically, we are interested in the way technology has impacted on communication and why specific solutions have been adopted above others. Understanding these issues, i.e. employing explanatory theories of change, will allow us to put into perspective the impact (and claims made about the impact) of digital information and communication technologies on the scientific article now and in the future.

In this final chapter, we first look at a number of theoretical considerations that help to understand the way technological change operates within the scientific community. We then focus on the digitization of the scientific article, and offer a number of more specific explanations for the discrepancy between expectations as found in the literature, and the findings of our study described in chapter 5.

1. Explaining the development of scientific communication

1.1 The closure of scientific communication

Although the scientific journal eventually became the primary vehicle for communication of the 'new science' in the 17th century, it has to be understood that this was a relatively slow development. As indicated above, it took almost a century for the scientific journal really to take off as a successful format. At its inception around 1665 there were a number of other formats in use by scientists. Nancy Fjällbrant (1997), in a discussion of the past and future of scientific communication, points to scientific books, newspapers, almanacs and calendars, and personal communication by means of letters, as well as to the use of coding systems such as anagrams for establishing priority of discovery.Fjällbrant explains the success of the scientific journal over these other forms by referring to Bijker's theory of the social construction of technology (SCOT).[1]

SCOT is a theory about the development of technology and the adoption of technological artefacts by society. Basically it states that the effects of a particular technology are not just a resultant of the intrinsic properties of the technology as such, but also depend on the social, cultural and political context in which the technology acquires its specific interpretation in terms of functionality and value. A fundamental idea in this theory is therefore that technological determinism is flawed: the impact of a technological innovation cannot be ascribed entirely and necessarily to its intrinsic technical properties.[2] Bijker introduces the concept of *interpretive flexibility:* the meaning of technology, and if and how it will be applied, are determined by whoever choose to use it and in what way. Initially, the technology may develop in different ways; different social groups apply different (and some times conflicting) interpretations based on their understanding of related technologies. This what is known as *technological framing.*

Furthermore, the properties of the technology are adapted within a specific technological frame in order to reduce problematic aspects. Eventually a single interpretation and related set of properties will emerge. This happens either because various groups reach consensus about the form and purpose of the technology, or because a single group is in a position to enforce its interpretation on others. Simultaneously, there may exist several competing technologies or technical artefacts that are suitable for performing the same function. Eventually one of these emerges as the dominant solution. The dominant solution is not necessarily the one that offers the best *technical* solution, but the one that best fulfills the needs, interests and interpretations of the various stakeholders involved.[3]

The concept of technological framing also points to the notion of frame incongruence as an explanation for many types of problems that arise in the course of introducing new technologies within a social system:

> 'Different technological frames imply different ways of knowing and making sense of technology. As these different interpretations are typically not articulated or discussed, they may [...] result unintendedly and unknowingly in misaligned expectations (such as technologists intending improvements in group work while users perceive improve-

[1] Bijker et al., 1987; Bijker 1995, 2001.

[2] Pinch and Bijker (1987, p. 22-23).

[3] The famous competition between the VHS, Betamax and Video2000 standards for video recording is a case in point.

> ments in individual productivity), contradictory actions (such as technologists installing and operating a technology while users wait for training and applications), and unanticipated organizational consequences (such as resistance, skepticism, and spotty adoption)'.[4]

'Social construction' points to the view that technology is not a neutral material artefact, but a social process in which meaning is attributed to technological phenomena and where they are interpreted with respect to their significance, potential use and usefulness, etc., but also with respect to diverse interests, ambitions and political aims. Collective framing at the level of a social group can be explained by *cognitive dissonance* at the level of the group. Cognitive dissonance[5] arises from a discrepancy between existing knowledge and assumptions on the one hand, and the perception of new, conflicting phenomena on the other. Due to the conflict, accommodation is required since the new phenomena cannot be assimilated, i.e. accepted without the need to change prevailing beliefs and practices.[6] Since accommodation involves the difficult process of changing beliefs and practices, it is often easier to re-interpret the new phenomena in a way that *reduces* the need for change, even if that defies the innovative properties of the underlying technology.

The process of social selection of a dominant technological solution and its interpretation is called *closure*, referring to the shift from a situation where many technical options remain open, to a closed situation where alternatives are no longer viable.[7] Once this has happened, the resulting solution is highly stable. Not only because there is agreement about the technical solution, but also because – as a result of that – significant economic, intellectual and organizational investments are made that impede the adoption of competing solutions. Nevertheless, closure need not be indefinite. It is often followed, after a certain period of time, by a process of *reopening*, where new options present themselves and/or are required by contextual changes. The social construction of technology is therefore in principle a long-term cyclical process with alternating periods of openness and closure.

Why did closure happen with respect to the scientific journal? Why did the article become the predominant form above the other available forms? If we look at the scientific journal as a technological artefact that shaped communication but itself was also shaped by the scientific community, we may conclude that three factors that were important to the various stakeholders (such

[4]Orlikowski and Gash 1994, p. 203-4.

[5]The theory of cognitive dissonance was formulated by Festinger (1957). For a discussion of cognitive dissonance in relation to the adoption of innovations, see Rogers 2003, p. 189-190.

[6]Atherton 2003.

[7]The term *closure* is also used by Thomas Kuhn to describe the emergence and adoption of a new paradigm in science (1996, p. 84). However, we do not wish to suggest that the mechanism behind the closure of media and other technological artefacts is identical to that behind scientific revolutions, even though both seem to share many characteristics.

as scientists, students, academic institutions and publishers) were involved in the closure of scientific communication: the need for more rapid dissemination, the need for open access to information without violating the intellectual rights of authors, and the development of an academic community with a distinct identity (Table 6.1). The development of group identity (culminating in the term 'scientist' first introduced by William Whewell in 1834) facilitated the convergence towards a common definition of the appropriate 'technology' for scientific communication. According to the SCOT-theory the emergence of a single group usually leads to the adoption of a more or less conventional solution.[8] It is therefore not surprising that the scientific journal (and its embedded literary genre: the article) evolved out of a conglomerate of existing forms rather than as a radical departure from tradition.

The scientific journal addressed these factors better than other publication forms, mainly due to three important properties:

- The adoption of *print*, primarily as a substitute for written correspondence between scientists that had been the predominant mode of rapid scientific communication.
- The choice of the journal *format*, allowing for a quasi-continuous flow of information in comparison with thebook form and providing contextualization through a multi-subject (if not often eclectic) confrontation of ideas and approaches. Also, the journal conveyed a sense of respectability and group identity to the burgeoning scientific community. In addition the journal format provided a convenient, standardized archival medium that allowed for referencing and long-term storage.[9]
- The introduction of various *control mechanisms* such as registration, peer review and bibliographical control, mainly as a result of the example set by the Royal Academy and its secretary Henry Oldenburg.

Bringing these elements together, we can now describe the closure of scientific communication as a process through which the scientific journal emerged as the embodiment of particular innovative properties that provided sufficient key benefits to the various stakeholders in a way that no other form (such as letters, printed books and newspapers) was able to provide (fig. 6.1).

What does the theory tell us about the current state of affairs with respect to the digitization of scientific communication? Many would argue that we are now entering a period of opening with respect to the scientific article as

[8]This is related to society's limited capacity for innovation and the the resulting requirement that innovation should develop along a single dimension rather than along multiple dimensions simultaneously.

[9]In addition, the periodical was a commercially attractive format for the publisher, requiring less capital investment and creating a more regular stream of income than other formats (Johns 2000, p. 162–163).

Table 6.1. Closure factors for the scientific journal

Speed	When the scientific revolution really took off in the middle of the 17th century the speed of invention and innovation increased significantly. This created a need for *rapid dissemination* of scientific information. This need was ultimately best met by the use of the printed journal as a communications medium.
Openness	One of the characteristics of the new science is that it is a developmental process: each new piece of knowledge is a building block that is put together with previous knowledge, and acts as a stepping-stone for further investigations. This process can only thrive if researchers share their knowledge by making it public, i.e. in a *state of openness*. One pre-condition of this is a mechanism that safeguards the '*intellectual ownership*' of research findings. Other pre-conditions are *quality control* (to prevent mis-information) and *archiving* (to ensure that information remains available and can be referred to). These pre-conditions were met by various control mechanisms that were embedded in the scientific journal system as it developed in the latter part of the 17th century.
Identity	During the 17th century scientists became aware of their belonging to a distinct social group, which created a need for *group identity*. This need was expressed by the creation of societies, the development of group norms and procedures, but was also enhanced by the scientific journal as a medium that was particular to the scientific community. Reading, but especially writing for these journals became a means for obtaining and expressing membership of this community.

the predominant form. It is certainly true that information technology seems to offer the possibility of new and competing forms. However, SCOT is an explanatory rather than a predictive theory. It helps us to analyze how and why a certain technological solution was adapted, and to understand how an ongoing process of technological development operates. But it does not readily allow us to predict the outcome.

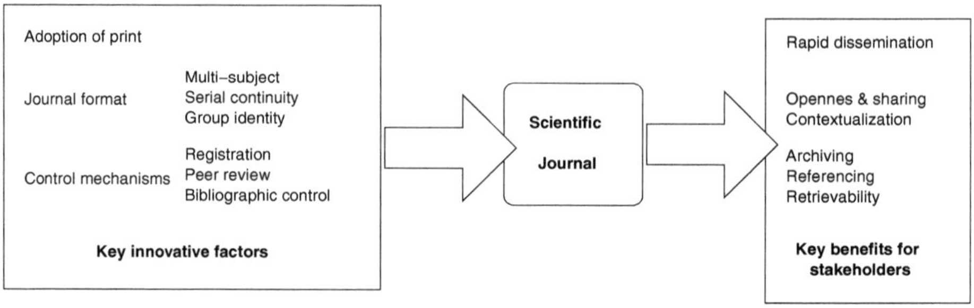

Figure 6.1. Closure and the scientific journal

Applying the concepts of SCOT to the current state of affairs, there appear to be two different technological frames, two distinct interpretations that prevail in two relatively distinct social groups.

One of these (Frame A) is the group of technological deterministic and optimistic, innovation-oriented activists that announce or wish to achieve a 'revolution' in scientific communication. This relatively small but vocal group includes both renowned scientists such as Harnad and Odlyzko, and non-scientists such as Okerson, Nentwich and others. Most members of this group are actively involved in the organization of scientific communication (e.g. as librarians, editors, publishers, researchers or consultants) and the development or application of new technological solutions in this domain. The group frames information technology as a potential or even inevitable source of change and improvement of scientific communication. As a corollary, this frame identifies significant defects in the current practice (often denoted as a 'in a crisis'), and denounces institutional actors (notably publishers) and 'non-believers' as unenlightened, conservative or even as enemies.[10] The group is joined by institutional managers (e.g. research directors, head librarians, faculty deans) who are concerned about the rising cost of scientific communication and therefore choose to frame technology as a more cost-effective solution.[11]

The second group (Frame B) consists of the large body of scientists who have a neutral or even sceptical attitude towards technology (at least within the context of scientific communication), and are mainly task/application-oriented. This group frames information technology as a secondary issue that only bears

[10]For instance, the commercialization of scholarly publishing in the sciences has been described by university librarians as 'at the core of the economic problem' of scientific communication (Branin and Case 1998, p. 478).

[11]See for instance Schulenburger 1998, 1999. The framing of technology for scientific communication resembles that of educational technology, including educational managers' reflex to interpret technology as a solution for financial problems and to demonize justified resistance to its adoption – see Werry 2001.

upon scientific communication to the extent that it supports the practice of research without corrupting its social functions (e.g. providing recognition and status) and moral values (e.g. maintaining the highest level of quality and integrity). This frame can be expected to interpret information technology as a development that preserves and extends existing functions and values rather than as an innovation that radically transforms a communicative practice that has evolved over the centuries in a way that is perceived as having reached a high degree of perfection. The attitude towards technological development within this frame is exemplified in a statement by Joshua Lederberg, president emeritus of Rockefeller University:

> 'A scientific publication is a grave act to be undertaken with the utmost seriousness [...] So, if we had never heard of the Scientific Journal in its print form, and were just watching the manifestation of this communication as it is operating today on the net, I think we would very quickly come to the conclusion we had better invent something like a peer review journal in order to provide some modicum of order and of discipline in that medium'.[12]

Although, as already stated, SCOT does not help to predict the outcome of the conflict between both frames, it does suggest that the second frame, interpreting information technology as secondary, supportive and extensional rather than as dominant and transformational, will not readily give way to the first frame as long as its vested power and traditions do not give way to significant external pressures.

1.2 An evolutionary view of scientific communication

Another approach to interpreting the development of scientific communication is provided by evolution theory. The term 'evolution' is often used to indicate any form of gradual change. But evolution in the scientific sense is a theory that not only refers to change, but also *accounts* for change. Science is not the responsibility of a single person but an activity that is carried out by a large group of people. So if an individual scientist changes his way of doing research, we do not say that the practice of research has changed. That is only the case if (a) the change becomes *common* practice and (b) it proves to be sustainable for a significant period of time (perhaps then being superseded by yet another practice). The idea behind this is that, within the context of a common practice, there will always be a certain amount of variation around the common standard. Usually these variations are relatively insignificant and/or short-lived. Sometimes they are significant, but do not seem to catch on, i.e. are not taken up and imitated by other members of the group. Sometimes, however, a significant variation, a substantial change in one or more properties of the common practice, does catch on and becomes part of the 'standard

[12] Lederberg 1996.

procedure' within the group for a substantial period of time. Such a change is not gradual and imperceptible, but measurable with respect to an identifiable property. This is the type of change that is referred to as 'evolutionary' and that is explained by evolution theory.[13] The necessary ingredients are a group with common characteristics and a certain amount of variation of these characteristics. These variations can be seen as responses to internal or external (i.e. environmental) pressures (adaption). At the same time, the extent to which variations are adopted as common characteristics or integrated into a new practice also depends on internal and external factors. The three concepts fundamental to this way of thinking are *innovation* (the introduction of new variants), *selection* (the adoption of one of the available variants by the group) and *reproduction* (the copying of the new practice by successive generations).

Evolution theory is, of course, in origin a *biological* theory explaining the development of biological species over time. It has, however, been adopted in various forms for explaining social and cultural process as well, and is then usually referred to as *selection theory.*[14]

A significant distinction between biological and cultural evolution or selection is that the former is based on random mutation, while the latter is believed to be teleological:

> '[selection theory] is a model which may be applied to other adaptive processes, or other apparently teleological series of events in which modifications seem guided by outcome'.[15]

In other words: the selections that are made in the process of social and cultural evolution are thought to be based on objectives (e.g. increased efficiency) rather than on chance.[16]

[13]Note that the *outcome* of an evolutionary process can often be described as a 'revolution'. For instance, Thomas Kuhn draws an analogy between Darwin's evolutionary theory of natural selection and his own theory of scientific revolutions (Kuhn 1996, p. 172).

[14]Although selection theory has its roots in a number of 19th century thinkers, the classic formulation is by Albert Galloway Keller (1915). See Fog 1999, especially chapter 2 for a historical overview. Evolutionary theory has been applied to conceptual change in science by Hull (1990, 2000) and to technological development by Nelson (1987).

[15]Campbell 1956, p. 330.

[16]One of the difficulties with selection theory arises from these teleological properties, as the theory itself offers no guidelines for deciding what the governing objectives are. Gross et al. argue that the governing objective of scientific communication (at least as far as the scientific article is concerned) has been efficiency of style and presentation, necessitated by the exponential growth of scientific periodicals since 1700 (Gross et al., 2002, p. 219). However, once we accept a relationship between volume and efficiency, the question arises why the objective of efficiency has not served as a limit to growth. Stylistics changes in scientific literature (e.g. a shift from 'conversational' to 'technical') can also be explained differently, e.g. as the outcome of increasing scientific controversy (cf. Latour 1987, chapter 1).

Selection theory and the scientific article

Selection theory has been used by Gross et al., to explain the evolution of the scientific article.[17] Their argument can be summarized as follows. Although each scientific article is different, they all show common characteristics. In evolutionary terms, each article is a distinct *phenotype*, while conforming to a common *genotype:*

> 'The *genotype* is the set of generative structures by which the organism inherits phenotypes and suites of characters. [...] In the case of the scientific article, the genotype is a set of predispositions: to create arguments, [...] to transform these arguments into sentences and paragraphs, and finally, to order these sentences and paragraphs according to well-organized organizational constraints. These predispositions are behavioral tendencies generally shaped by learning; [...] When the situation calls for it, scientists activate these predispositions to create a scientific article'.[18]

The idea is then that although particular manifestations of the genotype (i.e. individual articles) are relatively uniform, they will also show a certain degree of variation. These variations may be accidental, but are often due to environmental pressures. Selection operates on these variants in the extent to which they are reproduced, thus changing the 'species' (the genotype of the scientific article) over time: 'selection is the differential reproduction of variants as a consequence of environmental pressures'.[19] In the specific case of the scientific article, three environmental pressures provide a teleological basis for evolution, bearing on style, presentation and argumentation: the quest for efficiency, the quest for objectivity, and the internal debate about arguments (fig. 6.2).

Applying an approach based on selection theory to the future *digitization* of the scientific article leads to a certain degree of caution about a 'revolution' in the communicative practice of scientists. From the evidence of history we can conclude that scientific communication has evolved gradually, with a timescale to be measured in centuries rather than decades. The development of the scientific article (and of scientific communication in general) is an evolutionary process in which change is induced by external pressures, resulting in the selection of certain options above others. Scientific communication can be seen as an experimental pool, a social laboratory where all kinds of ideas and solutions for communicating scientific knowledge are proposed, interpreted and evaluated.[20] Some of these fit the requirements of science better than others, and therefore stand a better chance of being selected and adopted, often as a substitute for earlier solutions.

[17] See also p. 41 ff.

[18] p. 218.

[19] p. 219.

[20] The same applies, of course, to science itself. Cf. Gross et al.: 'In terms of selection theory, science advances ... by allowing a candidate swarm to compete in an environment in which only a few claims of fact and theory can survive.' (2002, p. 217).

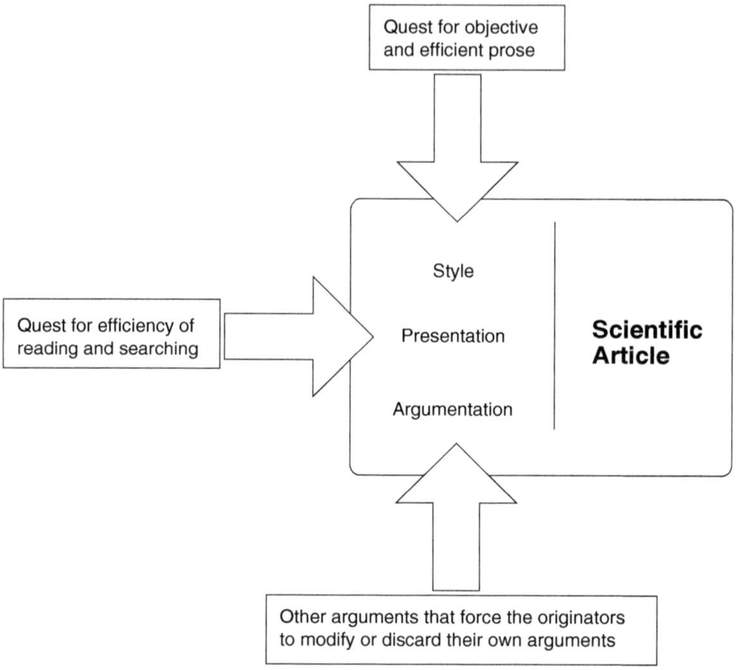

Figure 6.2. Evolutionary model of the scientific article

Some of the pressures bearing on scientific communication, generating variants and governing their selection are *internal* to science, concerned with changing demands of scientific disciplines and the research community. Technological developments and artefacts are but one of the many *external* pressures that influence the production and selection of variants in the social process of scientific communication (fig. 6.3). Other external pressures include those coming from – to give but a few examples – economic, legal and political domains. Without such pressures (based on a misfit of current practices plus the availability of better options), change will not take place. At the same time, responses to external pressures in the form of variations in the communicative practice have to come from *within* the system, generated by the scientific community itself. Therefore, change does not occur by the 'injection' of new technology. Rather, technology is adopted (or rejected) by scientists when it helps (or does not help) to resolve internal and external pressures. It seems probable that scientists will not and cannot change the style and presentation of their writing just because a new technology has become available. And when they do, changes can be expected to be minor and tentative.

The outcome of evolutionary development processes such as that studied here is emergent and contingent. It is emergent in the sense that the outcome of

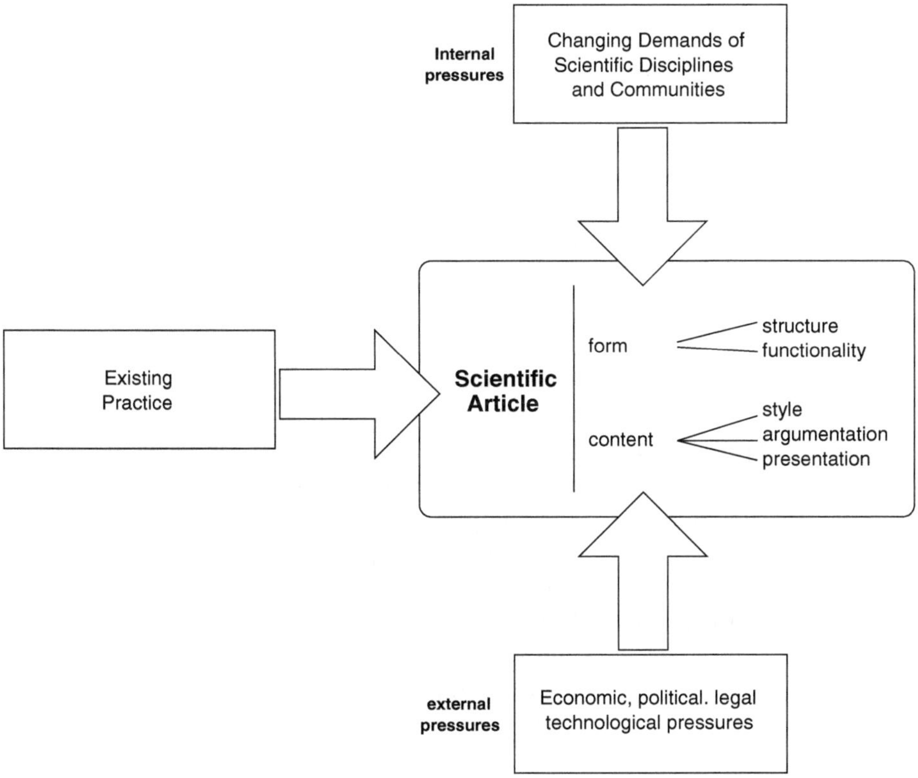

Figure 6.3. Evolutionary pressures on the scientific article

a change process cannot be deduced from the pressures that bear on it, however relevant these pressures are to the outcome. And it is contingent in the sense that the outcomes can be expected to fit the pragmatic context of the actors (e.g. scientists) rather than any theoretical or ideological model. When closure has happened, the process and its outcomes can be described in a logical fashion. But as the result of a highly complicated process with a multitude of pressures bearing on the path towards closure, the outcome, however logical once it has been achieved, cannot be predicted from the beginning.

1.3 The diffusion of innovations

A third way to conceptualize technology-induced change is in terms of innovation theory. We have used this approach in an early exploratory study of

Table 6.2. The process of technological innovation

Procedures	The way a certain application or method is carried out, often expressed as a specification, protocol or standard (e.g. the tcp/ip protocol specifies how messages are transported over a network).
Components	The artefacts that result from the application of one or more procedures, e.g. a cdrom or an USB memory device.
Systems	Functional, self-contained combinations of a larger number of components, e.g. a personal computer or MP3-player.
Applications	The employment of systems for a specific purpose, e.g. document processing, file storage.
Infrastructures	The complex of actors, functions and facilities necessary for an application to have a structural impact on a segment of society.
Innovations	The large-scale adoption of infrastructures, resulting in general acceptance and common usage of applications as a substitute for or complementary to existing applications.

the scientific and technical information infrastructure.[21] We argued there that any technological innovation is the outcome of a *process* in which standardized procedures are transformed into technical artefacts, which then acquire a specific functional interpretation in the form of an application; applications then aggregate into technical infrastructures (table 6.2).[22]

An infrastructure is a complex, multi-valued notion that has been described as a 'relational property':[23] the infrastructure exists in the binding of con-

[21]Mackenzie Owen and van Halm 1987, 1.2.1-1.2.30, Mackenzie Owen and van Halm 1989, p. 8-34.

[22]These stages need not be sequential in the sense that the development of technology traverses them in the described sequence. Although a given level (e.g. an application) requires and can be reduced to 'lower' levels (e.g. systems, components and procedures), the development of these lower levels continues and may contribute to higher levels at any time. The various levels, notably systems, applications and infrastructures, are social constructs based on a specific interpretation of technology and the functions it is supposed to perform. For instance, the personal computer has changed considerably over the years, and has incorporated and benefited from the continuing development of procedures and components. Its functions (the things people use it for) have also changed considerably, e.g. evolving from word and data processor to network communication device to multimedia machine. Nevertheless its interpretation as a specific type of system, the 'pc,' has remained virtually unchanged.

[23]Star and Ruhleder 1996.

stituents rather than in the set of constituents themselves. Infrastructures are not only the outcome of a social process involving the coming together of disciplinary heterogeneous actors, technically, politically, economically and culturally motivated decisions, the development and application of classifications and standards, etc. (as is also the case with the underlying levels such as applications, systems and components). They also shape (and to a certain extent define) social reality in the way they structure, define and regulate social (inter)action. As Bowker and Star argue, one of the characteristics of infrastructures is that under normal conditions they become invisible. We still see our pc as a 'thing' on our desktop, but the complex infrastructure that renders it a *useful* thing escapes our attention: the infrastructure only becomes visible when it fails.[24] Younger people who have grown up with the pc hardly notice it at all, and are no longer aware of its technical properties such as speed, capacity etc. For them the pc has dissolved in the general information infrastructure and has become a 'normal' and hence transparent component of their day-to-day lives.

If these infrastructures do not receive wide-scale adoption as the basis for localized practices of use, the innovation ultimately fails. In other words: the infrastructure *is* the innovation. The innovation is not defined by the technological artefact or invention, but by its wide-spread adoption leading to a fundamental change in the organization of the social context in which it functions. We might therefore equate the *process* of innovation with the emergence of an infrastructure and its acceptance in terms of its disappearing from the awareness of users. Technology is innovative precisely when it becomes transparent.

Because innovation implies a *change* process within an organizational or wider social context, it is useful to distinguish between three types of organizational change:[25]

- *first order change* is an incremental process that modifies the existing situation based on known problems and solutions;
- *second order change* marks a transformation, i.e. a radical and discontinuous shift from an existing stable situation to a new stable situation via an unstable phase where problems need to be identified and agreed on, and solutions are often not evident;
- *third order change* is an intervention that does not necessarily change the current situation, but rather the organization's or social group's capacity for change, e.g. in preparation for a situation of 'continuous' change. The main change realized here is a distancing from traditional cultures, processes and frames of mind.

[24] Bowker and Star 1999, p 33–37.
[25] Bartunek and Moch 1987; Orlikowski and Gash 1992.

It is probably true to say that most innovative processes (especially when they are intentional and policy-based) involve incremental, first order change. 'Revolutionary' processes governed by second order change are relatively rare, perhaps because they would seem to require a preceding phase of third-order change to be successful.

The underlying causation of innovations is not always the same. Some innovations develop incrementally, driven by competition between independent economic agents. Other innovations are the result of major scientific breakthroughs leading to a series of applications until they are superseded by new breakthroughs.[26] Information and communication technologies are an example of a third, convergent type of innovation where separate technologies (e.g. computing, communications, electronics etc.) that existed before come together and achieve an innovative force that exceeds that of the individual components.[27]

The process of technological innovation is usually described in terms of *diffusion*, defined as 'the process in which an innovation is communicated through certain channels over time among the members of a social system'.[28] The accepted viewpoint is that innovation is a process that is based on social interaction, i.e. on a process of communication in which individuals converge towards a common interpretation (in terms of nature, value, desirability, applicability etc.) of a new idea, with as the outcome a change in the behavior of the (majority of the) participants. While approaches based on the social construction of technology focus on the way technology is shaped by social processes, diffusion theory aims at explaining the *adoption* of new ideas and artefacts within a social group or system.

Basically, diffusion theory tries to describe in quantitative terms the degree of adoption over time, and the relationship between internal and external factors and the level of innovation.[29] The nature of an innovation diffusion depends on this relationship between internal and external influences. The diffusion of ICT within scientific communication is typically a case of mixed (internal and external) influences. Internal influences include the role of scientists

[26]In other words: technological revolutions tend to follow scientific revolutions or paradigm shifts (Kuhn 1996).

[27]McClelland 1994, quoted in Stout 1999, p. 334.

[28]Rogers 2003, p. 5. Rogers' book on the diffusion of innovations (in its second edition published as 'Communication of innovations') has been the *locus classicus* of innovation theory since its first edition in 1962.

[29]See Mahajan and Peterson 1985 who provide a mathematical formulation of the *fundamental diffusion model* as $\frac{d\,N(t)}{dt} = g(t)\,[\overline{N} - N(t)]$, where $N(t)$ denotes the cumulative number of adopters at time t, $\overline{N}$ the total number of *potential* adopters, $\frac{d\,N(t)}{dt}$ the diffusion rate and $g(t)$ the diffusion coefficient. When $g(t)$ is independent of $N(t)$ the diffusion is governed by *external* influences, and when $g(t)$ is totally dependent on $N(t)$ it is governed by *internal* influences. Therefore, $g(t) = (a + b\,N(t))$ denotes the *mixed* influence case.

and the research community in adopting new ideas, methods and artefacts. External influences include the role of more peripheral actors such as publishers, libraries, and academic management (notably acting as change agents, see below), as well as more general factors such as economics, legal considerations, politics etc.

The fundamental model of diffusion of innovations is based on the assumption that diffusion is *binary*, i.e. the innovation is either adopted in full or not at all. In the case of the digitization of scientific communication, however, the innovation may take on many different forms and be applied to different degrees. Even full acceptance of digitization by the academic community would therefore imply an approximately normal probability distribution (or an Aristotelian 'golden mean') of the *extent* to which digitization is used. In other words: relatively few scientists would adopt all (or none) of the possibilities of digitization, with a larger group adopting digitization 'to a certain extent'.[30]

The diffusion of innovation is a complex process that is governed by a multitude of determinants. These include properties of the innovation itself, the way *decisions* about adopting the innovation come about, the nature of the *communication channels* involved, the nature of the *social system* that is to adopt the innovation, and the role of *change agents.* The various properties (as perceived by members of a social system) of the innovation are defined by Rogers as *relative advantage*, *compatibility*, *complexity*, *trialibility* and *observability.* Of these, all but complexity are positively related to the rate of adoption of the innovation. For instance, the more compatible an innovation is with what a potential adopter already knows, believes and uses, the easier it will be adopted.

We now discuss these determinants in the context of the digitization of scientific communication.

- **Innovation attributes**

  – *Relative advantage.* The relative advantage of digitization must be related to possible improvements in the system of scientific communication. As Gross et al. have pointed out, the scientific article has achieved a high degree of perfection as a rhetorical genre for communicating research ideas and results. Scientists themselves do not seem overly critical towards the genre. Advantages, if any, would have to be perceived in the areas of presentation (i.e. the organization and visualization of materials) and enhanced argumentational devices. At

[30] Stated differently, this implies that the frequently used S-shaped logistic diffusion curve (or the Gompertz function $\frac{d\,N(t)}{dt} = b\,N(t)\,\left[\ln \overline{N} - \ln N(t)\right]$) for describing the adoption of innovation is misleading in that it suggests a total level of adoption that will never be reached if all participants adopt the innovation, but only to a certain extent. On other words, any point on the Gompertz curve denotes not a unary value but a set of (normally distributed) values.

the aggregate level (i.e. in terms of scientific journals and the overall infrastructure for scientific communication) more concerns have been expressed. These can be summarized as: high cost, delays in the publication process, scattering of papers due to a proliferation of journals, and problems with the refereeing system (e.g. reviewer bias).[31] Advantages over the current system in these areas can be expected to be a positive force for the diffusion of digitization.

- *Compatibility.* This factor is related to relative advantage, and simply refers to the fact that change generally provokes resistance. This implies that the use of easily adopted forms of digitization that involve little or no change to work practices is to be expected rather than the adoption of more advanced forms. Compatibility will be less of an issue to the extent that scientists are familiar with other forms of digitization and with information technology in general.
- *Complexity.* The development of the scientific article can be seen as a process of complexity reduction, i.e. an evolution towards a limited repertoire of standardized communication channels and rhetorical devices.[32] This can be explained from the fact that research rather than communication is the primary concern of scientists, and that communication should therefore not distract from the more important research activities. This suggests that any added complexity due to digitization is perceived as a barrier to research and can therefore be expected to influence diffusion negatively. A corollary of this is that the use of digitization by the individual author is less likely to be adopted than digitization at the level of the infrastructure, where the burden of complexity is primarily carried by external actors. However, since scientists are also users of the information infrastructure (especially in their roles as information seekers and readers), complexity remains an issue here as well.When digitization reduces complexity both for the author and for the reader, its chances of adoption can be expected to increase significantly.
- *Trialibility.* Social constraints (e.g. related to peer review, recognition and status) argue against a high degree of triability: for the individual there is little to be gained (and potentially much to be lost) from experimenting with digital forms in a context where traditional forms are the norm. However, the Internet does offer scientists many other contexts (such as personal websites or e-conferences) where it is easier to

[31] Kim 2001, p. 37.

[32] The continuous specialization of scientific domains (and of scientific journals) can also be seen as a way of coping with increasing complexity by reducing the domain of discourse to a more limited but manageable scale.

experiment with digital forms. If successful, these may gradually spill over to the formal context of peer reviewed publications.

- **Observability.** It is important for the diffusion of innovations that they are visible to potential adopters. In the case of the scientific article, this would mean that new digital forms would have to have a significant presence in the array of resources consulted by scientists. In view of the rather limited number of 'E-only' journals, this is clearly not the case as far as innovative digital forms of the scientific article are concerned. However, here too there may be a spill-over effect from other forms of communication.

- **Innovation decisions.** The diffusion of innovations is governed by the decisions of individuals to adopt or reject the innovation. This type of decision is a process that evolves through several stages usually referred to as knowledge, persuasion, decision, implementation and confirmation. At any stage the innovation may be rejected. A number of observations can be made pertaining to this decision process in the context of scientific communication. One is that the individual scientist is not in a position to make a free decision. Constraints predicated on the nature of the social system (see below) and policy rules imposed by publishers and editorial boards limit the range of options for adopting innovations. Another observation is that there is a relationship between awareness and need in the process of decision making. Research does not provide a clear answer as to which of these precedes the other.[33] But in view of the previous observation it would seem unlikely that mere awareness is sufficient to move scientists to adopting an innovation that would significantly change a well-established practice. It is likely, therefore, that the adoption of digitization by the scientific community would have to be grounded in a widespread need for change towards perceived advantages of the technology.

- **Communication channels.** Discussion of the role of communication channels in the diffusion of an innovation that itself can be seen as a communication channel is complicated. What is at issue here are the channels through which potential adopters are informed of the innovation. Two very different channels are involved in the innovation of scientific communication. One is the discourse of change agents and supporters of new forms of scientific communication. Although this discourse is carried out through 'normal' channels such as journal articles and conferences, its audience appears to be rather restricted and its activities limited to 'preaching to the converted'. The other channel is the scientific journal itself, that serves not only as the scientist's vehicle for communication, but also as its example. The role of

[33] Rogers 2003, p. 172.

the journal article as the 'canonical form for the communication of original scientific results' makes it an extremely powerful medium for diffusion of its own innovation. By its very nature, therefore, the natural channel for innovation of the scientific article is the scientific article itself.

- **Nature of the social system.** Scientific communities are relatively closed social systems with long-established norms and practices.[34] They lack centralized decision-making mechanisms. Members have a professionally critical attitude towards new ideas and a need for a 'proof of concept' as well as a sense of necessity and inevitability before changing their work practices. Various studies show that there is a lack of new agreed social norms pertaining to electronic publishing, and that there is uncertainty amongst scientists about the stability, reliability and quality of electronic publications and whether electronic publishing will provide the same amount of recognition as print publishing.[35] This leads to the expectation that scientists will not easily embrace digitization of their communicative practice unless other factors are particularly favourable.

- **Change agents.** A change agent is an individual or group capable of influencing the members of a social system to adopt an innovation and change their behavior in a certain direction. Internal change agents (i.e. members of the social system itself) are also called opinion leaders, and can be used by external change agents to achieve their objectives. An innovation *champion* is a 'charismatic individual who throws his or her weight behind an innovation, thus overcoming indifference or resistance that the new idea may provoke in an organization'.[36] In scientific communication the agents for change towards digitization are mainly external to the scientific community proper. The need for opinion leaders and champions is manifest in the efforts of the Public Library of Science not only to attract 'big names' (such as Nobel laureates Harold Varmus and James Watson) as contributors, but especially as publicity agents, e.g. figuring on posters, TV spots and other promotional materials.[37]

In small countries the conditions for infrastructural technological innovation are easier to fulfill than in large countries, and therefore the international (or 'global') character of scientific and technical communication can be seen as a barrier to innovation.[38] A theoretical underpinning for this is found in another

[34]Cf. Whitley 2000. See also Thomas Kuhn's description of 'normal' science (Kuhn 1996, ch. 2-4).
[35]Zhang 2001. A study by MIT Press showed that scientists are more likely to submit papers to an electronic journal if it also has a print version (Kiernan 1999).
[36]Rogers 2003, p. 414. Rogers cites Donald Schön (1963, p. 84) in his emphasis in the importance of champions: 'The new idea either finds a champion or dies'.
[37]See http://www.plos.org/support/stuff.html.
[38]Mackenzie Owen and van Halm 1989, p. 17. See also Bruland 2001.

of the assumptions of the fundamental diffusion model that (at least when the influence is internal or mixed) the members of a social system mix completely, i.e. that there is full 2-way interaction between prior and potential adopters. This is clearly not the case in scientific communication for three reasons. First, the geographical spread of scientists is global, and interaction is therefore necessarily limited. Second, science consist of a large (and increasing) number of social subsystems with limited mutual interaction. Third, when there is interaction between scientists, they discuss their research rather than their mode of communication. This suggests that science in general, as an international enterprise, might not easily yield to radical changes induced by new technological developments. Such changes could, however, manifest themselves more easily in smaller disciplinary domains. This localization might provide at least a partial explanation for the fact that a substantial innovation such as the wide-scale use of pre-print servers has at least initially been limited to a modest number of smaller academic fields such as high-energy physics and cognitive science.[39]

Theory also suggests that the diffusion of digitization might proceed more rapidly in the infrastructural domain than in the domain of scientists as authors. The adoption of new communication modes by scientists as authors is restricted by a lack of interactions, which also tends to fixate social restrictions such as received norms, competition and the need for recognition by peers.[40] On the other hand, the group pressing for change (cf. 'Frame A', p. 196) is to a large extent coherent, well-organized, and related to infrastructural responsibilities (e.g. libraries, management, open access publishing) rather than involved in research and authorship.

To summarize our discussion of the determinants of diffusion, we find that successful digitization would have to address the perceived problems of cost, publication delays, scattering and peer review. It will be most successful if it does not involve significant changes in the work practices of scientists or add to the complexity of the communication process. Experience with new communication forms outside the realm of peer-reviewed publishing might have a positive spill-over effect on the digitization of formal communication. The nature of the social system and the prevalence of external change agents are negative factors for innovation of formal communication genres. The chances for adoption of innovations are highest at the level of the communications infrastructure, and relatively low at the level of the communicative practice of individual scientists.

[39] E.g. the arXiv.org e-Print archive (http://arxiv.org/) for physics (Ginsparg 2001) and the Cognitive Science E-prints Archive (http://cogprints.ecs.soton.ac.uk/). For an overview of current (pr)e-print servers see http://www.osti.gov/eprints/colldesc.html.

[40] See the description of 'Frame' B on page 196 and the reference to Whitley below on page 212.

2. The myth of the technological revolution

The ‘revolution’ in scientific communication that is supposed to be caused by information and communication technologies has often been compared to the so-called ‘Gutenberg revolution’.[41] But as we have seen, that revolution is more myth than reality as far as science and the media of scientific communication are concerned. The scientific journal was not a direct result of the invention of printing, but was predicated on a transformation – over a period of two centuries and more – of science itself that was facilitated rather than caused by the printing press.

The effects of the printing press are to be described more in terms of distribution (to a larger and wider audience), authorship (including the ‘entrepreneurial’ writer) and content (a wider and increasingly non religious and non-scholarly subject matter) than in terms of the medium itself. Many significant properties of the modern system for scientific communication that developed from the latter part of the 17th century onwards were already in place in the early universities. The scientific journal itself subsumed many characteristics of other forms, and closure towards a the standard form of scientific journal and its embedded article took considerable time.[42]

As we have seen, it took considerable time before the journal established itself as the primary medium for scientific communication (especially at the expense of the monograph), slowly evolving towards the standardized form that we have come to know in the 20th century. Kronick writes in his dissertation on the history of scientific and technical periodicals that ‘technology itself had very little impact on the periodical’s production and distribution. The processes of printing, paper-making and transportation remained remarkably stable throughout this entire period’.[43]

Although printed media certainly facilitated the development towards modern science, it would be an exaggeration even to suggest that the Scientific Revolution was caused by the new medium. History teaches us that media do not change the practice of scientific communication let alone of science itself. As the practice of research develops and the requirements for communication change, appropriate media are chosen (when available) or developed (when not). Closure follows practice, not the other way round.

Our analysis of various theoretical approaches to technological change leads to a number of conclusions:

[41] Harnad 1991; Birkerts 1994; Füssel 2001; Giles 1996; Hammes 2001; Siler 2000.

[42] Johns 2000.

[43] Kronick 1976, p. 47-48.

- The adoption of technological solutions is based on social interpretation, sense making and acceptance leading to closure and the dominant position of one particular solution with respect to other available solutions.
- Closure is a relatively slow, evolutionary process of accommodation and adaptation to external factors.
- Innovation diffusion processes follow a distinct development path, leading either to widespread adoption or to rejection and failure.
- Technological development is also a political process where change agents and power relationships play an important role.
- The success of a technology depends only to a limited extent on its technical properties and merits. Many other factors contribute to success or failure. Both history and the various theories of technology-related change teach us that using the perceived merits and potentials of new technology to predict its future use and impact is not a good idea.

Many accounts of technological development and the impact of 'new' technologies (in whatever historical context) view technology as an abstract construct that exerts an influence on its environment and acts as an agent (if not *the* agent) of change. Such accounts tend to describe the outcomes of these influences in terms of 'revolutions': fundamental changes that necessarily derive and can be explained from intrinsic properties of the technology. These accounts are essentially predicated on technological determinism and fail to consider the manifold interactions and interdependencies between technology and social context. All too often, then, these accounts lead to false ideas about the inevitability of technological outcomes and the 'logic' of its consequences for society, and to activism aimed at promoting inescapable developments and pursuing a battle against more 'conservative' and 'ignorant' forces.[44]

As we have seen, theoretical approaches to technology and technology-related change nowadays tend to take a more balanced view and to describe social contexts as agents of technological change rather than the other way round. The stages in the process of technological innovation that we have described earlier (see table 6.2) can be viewed in terms of increasing contextualization: whereas procedures and components are to a large extent technologically determined, their application and embodiment in infrastructures – and therefore their large scale adoption – are determined by social contexts. These various approaches that go beyond the technical properties of technology can be subsumed under the broader theory of the *social shaping of technology* (SST):[45]

[44]For an extensive critique of the utopian rhetoric surrounding information and communication technologies in general, see Robins and Webster 1999.
[45]Williams and Edge 1996; Williams 1997.

'We argue that a variety of scholars, with differing concerns and intellectual traditions, find a meeting point in the SST project. They are united by an insistence that the 'black-box' of technology must be opened, to allow the socio-economic patterns embedded in both the content of technologies and the processes of innovation to be exposed and analyzed. [...] SST studies show that technology does not develop according to an inner technical logic but is instead a social product, patterned by the conditions of its creation and use. Every stage in the generation and implementation of new technologies involves a set of choices between different technical options. Alongside narrowly 'technical' considerations, a range of 'social' factors affect which options are selected – thus influencing the content of technologies, and their social implications'.[46]

Williams also argues that the *configurability* of new media technologies allows them to be adapted to specific social contexts. This happens in a way that could not be achieved by mechanical technologies that lacked this degree of configurability, such as the printing press. Therefore the printing press had to be a shaping force if it was to be a force at all, whereas new information media are much more susceptible to *being shaped*, accommodated and used in many different ways according to context. Within the context of scientific communication, one would expect the result of digitization and the use of new technologies to differ between different disciplinary fields, a point made by Eason et al. and Kling & McGim, and also corroborated by Nentwich.[47] Designs aimed at engineering a new system for scientific communication based on a monolithic conception of information technology fail to take such field differences into account.[48]

The 'revolutionary' impact of digitization that informs many accounts of the current and future state of scientific communication is not supported by theories of technological development. References to the 'Gutenberg revolution' suggest a direct and inescapable link between technological inventions and the practice of scientific communication. That link does not exist. The more revolutionary accounts seem to stem from the same type of myopia described by Eisenstein with regard to the role of the printing press.[49] New developments tend to have a 'magnifying effect', drawing attention away from what is so ubiquitous as to have become invisible. In the case of current developments it is not only a distorted view of history – reflected in references to the 'Gutenberg revolution' – that plays a role, but also a misunderstanding of the significance of more traditional forms of scientific communication for the research context that in itself is highly competitive and innovative. In addition, revolutionary accounts of new media underrate the role of social factors in the adoption of new technology.

[46] Williams and Edge, p.866.

[47] Eason et al., 1997;Kling and McKim 2000; Nentwich (2003, chapter 3).

[48] See for instance Buck et al., 1999; Smith 1999a; Hurd 2000; Gass 2001; Hammes 2001; Van de Sompel et al., 2004. For a more balanced view see Atkinson 2000.

[49] Eisenstein 1980, p. 17, 39.

Whitley describes the modern sciences as 'reputational systems of work organization and control' where the communication system is 'the major agency of social control of competence standards and work process as well as being the locus of negotiations over intellectual goals and priorities'.[50] In a context where scientists compete for recognition and where a high level of task uncertainty entails both opportunities and risks for participants, it is inevitable that the communication system itself is seen and welcomed as a stabilizing factor. Scientists who are required to compete on the basis of their research procedures and outcomes cannot be expected to have a great desire to distinguish themselves on the basis of their mode of communication. Where reputation and career are continuously at stake, scientists will tend to prefer established modes of communication that do not draw attention away from the significance of their research itself. In other words: the nature of scientific work brings about a culture where established communicative norms and conventions prevail because they are functional and supportive. This perforce creates a certain degree of resistance to change. Innovations will only be adopted if they (a) benefit both authors and readers and (b) do not violate other norms and conventions of the scientific community. Innovations that do not affect the style, presentation and argumentation of the author, but rather pertain to the distribution mode at the infrastructural level can be expected to be more acceptable to the academic community.[51]

These theoretical considerations are supported by the current practice of scientific communication. New technological developments in the area of ICT have sparked off innovative approaches, but mostly in the realm of *informal* communication (e-mail lists, portals etc.), derivative information products (e.g. virtual journals) and, especially, data sources operative at the input phase of research.[52]

At this stage of our study the main impact of ICT on *formal* scientific communication through the peer reviewed research article seems to be in the area of the transmission mode (i.e. via the network rather than in print). The vast majority of what scientists perceive as 'e-journals' are digital versions of (and at the level of the article exact copies of) existing journals that are (or at least have been) available in printed form. Only a limited number of 'e-only' journals exist as the outcome of the first phase of digitization. This is precisely the

[50] Whitley 2000, p. 34.

[51] It is illustrative in this context that an Ad Hoc Committee on the future of publishing of the Modern Language Association included in its final report a number of recommendations for university departments, libraries, publishers and university administrations, but not for scientific authors. See MLA Ad Hoc Committee on the Future of Scholarly Publishing 2003.

[52] A bibliometric analysis by Kaminer and Braunstein (1998) has demonstrated that Internet use has a significant positive effect on the productivity of scientists. Most of this use is attributable to informal communication. Only a fraction (9%) of the use of the Internet was related to gathering formal scientific information through e-journals.

stage where the new medium had the potential to evolve towards new modes of presentation and argumentation, irrespective of entrenched forms derived from the world of print. Our analysis of the way scientific communication has evolved over the centuries as well as the various theories of change described in this chapter suggest that this potential is unlikely to be realized.

3. Transforming scientific communication

Are there specific reasons why digitization has had less impact on the peer-reviewed scientific article than many would have expected? Why is it that, as we have noted above, scientists are eager to use the scientific journal in its new digital form, when they are reluctant to change their habits as far as creating their articles is concerned? Why has the revolution that was supposed to transform scientific writing not come about? In the previous sections we argued, on the basis of a number of theoretical approaches including the social construction of technology, evolution theory and innovation theory, that such a revolution is unlikely. In this section we shall discuss a number of more specific theoretical positions that can help to understand this issue.

3.1 The significance of the electronic journal

There exists a significant literature on the impact of ICT on scientific communication in general, and more particularly on the development, use and impact of e-journals.[53] However, most studies focus either on how and how much scientists access and read electronic scientific literature, or on publishing practices and scientists' attitudes towards new communication channels and publishing models. These studies address what we have described as the aggregate level of scientific communication. Very little attention has been given to the impact of ICT on the communication practice of scientists at the level of the individual scientific article. On the one hand, as we have seen in chapters 1 and 5, proponents of ICT and some journal editors envisage a transformation of scholarly writing through the use of digital techniques. On the other hand, many editors, publishers and authors seem to regard the use of ICT as neutral with respect to the scientific article: as a change of the distribution medium and access facilities that does not impinge on the genre of the scientific article itself.

As we have seen in the previous chapter, the impact of digitization on scientific articles themselves, as published in both the early e-only journals and the more recent open access journals, is indeed limited. It is also clear that by far the most journals available in digital form are just digital copies of their printed counterparts. This leads to the conclusion that the outcome of the process of

[53]For recent overviews, see Houghton et al., 2003, 2004; Tenopir 2003.

digitization since the late 1980's is a transformation of the *process* of scientific communication rather than of the *substance*. Distribution and access (including searching and browsing) of scientific information have become faster and easier. Various functionalities at the aggregate level of the journal, publisher or library facilitate the communication process. But the nature of the scientific article as the unit of communication has not changed significantly as a result of digitization, in spite of various attempts at creating innovative e-only journals that were expected to transform the scientific article into a new digital genre. Although digitization of the scientific journal has changed the way scientists acquire information in what we have described in chapter 3 as the input phase of research, it has not fundamentally changed the way scientists report on their research outcomes in the output phase. This finding is in line with our expectation based on innovation theory that in the case of scientific communication, innovation tends to be more successful at the level of infrastructures (i.e. what we have described as the aggregate level) than at the level of the communicative practice of individual scientists (see page 209).

This view is supported by various studies that indicate that e-journals are quickly being accepted by the scientific community, and this is predominantly due to their accessibility ('anywhere, anytime') and to the added functionality at the aggregate level.[54] There are also indications that digitization of the research journal is leading to changes in research practices, for instance in terms of an increase in multidisciplinary research, which can be explained by the fact that subject searches in electronic journal collections tend to cross disciplinary boundaries more easily than when searching print collections.[55]

Another important advantage of the e-journal for scientists is to be found not in what digitization adds to the structure and content of the article, but in the e-journal's 'intertextuality', i.e. in what digitization adds 'between' articles (and between articles and other information resources) in terms of active hyperlinks and other functionalities. In information science the document is often regarded as a carrier of knowledge. Information systems are set the task not only to select 'relevant' *documents*, but preferably to retrieve and extract relevant *knowledge* from these documents. However, as Hjørland (1998, p. 616) points out, 'people are not only seeking raw facts, but also substantiated knowledge claims. Therefore users are interested in background information, and this is transferred via documents/texts, informal communication and other means.' In other words, the information system has to provide not just content, but also context. Providing this context is exactly what happens at the aggregate level by means of search functions, cross-linking etc. This explains the importance of the embeddedness of the scientific journal in a networked

[54] Rusch-Feja and Siebeky 1999; Voorbij 2005, ch. 7.4
[55] Voorbij 2005, ch. 7.5.

context that links the article - acting as an entry node to the networked information system of scholarly communication - to a rich informational background.[56]

Our analysis points to an 'error of logical categories'[57] in much of the discussion about the electronic journal where the distinction between the aggregate level of the electronic journal and the specific level of the peer reviewed article is often not made. These discussions treat scientific communication as if it were a single logical category. But this is not the case. The concept of scientific communication has to be thought of as encompassing two distinct categories. One is the communication channel, whether described as a conduit system (consisting of one or more types of communication forums, e.g. the e-journal) that we have described here as the information chain or as a transaction space or continuum. The other is the specific genre (e.g. the scientific article) that is used as the vehicle for representing scientific knowledge and the outcome of research. As we have seen, the genre of the scientific article remains relatively unchanged even when the information chain as a conduit has been transformed more or less completely to the digital format. The conclusion that digitization of one category must inevitably include digitization of the other, is a mistake.

3.2 The illusion of new media

At the root of many claims for a transformational or revolutionary effect of the digital format on scientific communication lies a specific view of what digital media are about. This view regards the digital format as a 'new' medium, a thing of itself, distinctive from other media with its own 'identity' and specific properties, and therefore by nature different from other (and especially earlier) media. There is therefore a dichotomy between the old and the new medium, and modes of communication (e.g. communicating research results) are seen as shifting from one medium to the other. Furthermore, a positivist outlook on technological development regards the specific properties of the new medium as an advancement, and they are believed to offer new possibilities and to liberate representation from the 'restrictions' of earlier media. In this view, it is seen as inevitable that a mode of communication (e.g. formal scientific communication) will eventually put aside the old medium to adopt the new. Although it may cling to its traditional characteristics initially, it will soon be liberated and adopt the new characteristics. This is then believed to be sufficient justification for expecting a revolution or at least a transformation of the communicative practice as a consequence of the new medium.

[56] I.e. not just related documents from the same disciplinary domain, but also information from other domains, practice information, individual and institutional actors, background data, software tools, etc.

[57] Ryle 1963, p. 17 ff.

A related, but even more radical view is offered by Bolter and Grusin (1999) who anthropomorphize (new) media as agents that challenge and take over from existing media through a strategy of 'remediation':

> 'We will argue that these new media are doing exactly what their predecessors have done: presenting themselves as refashioned and improved versions of other media. Digital media can best be understood through the ways in which they honor, rival and revise linear-perspective painting, photography, film, television and print. No medium today ... seems to do its cultural work in isolation from other media... What is new about new media comes from the particular ways in which they refashion older media and the ways in which older media refashion themselves to answer the challenges of the new media.' (p. 14-15)
>
> '[We] call the representation of one medium in another *remediation*, and we will argue that remediation is a defining characteristic of the new digital media.' (p. 45)

This view represents media almost as autonomous actors that appear on the cultural scene more or less suddenly and autonomously, and to which, if they are successful, other media ultimately have to adapt. In this view existing modes of communication do not adopt the new medium, but it is the new medium that actively 'takes over' and transforms existing modes of communication.

As such these views of media as distinct entities are in line with an evolutionary position where variant manifestations of representation compete for survival: if a new medium has sufficient advantages over its predecessor, it will survive and replace other media. This position is understandable in view of the fact that evolutionary forms are compelled to adapt to changing environments. Without doubt, the shift towards digital environments in general can be expected to produce adaptive forms in the specific case of scientific communication, and many new, network-based scientific communication forums provide examples of this.[58]

However, these lines of thinking do not help to explain the digitization of formal scientific communication as described in this study. The transformation of the scientific article has not happened to the extent that might be expected on the basis of the intrinsic properties of the digital medium and that has been predicted in many writings on the subject over the years.

We would like to explain this by proposing a different view based on our discussion of innovation and the social construction of technology in chapter 2. There we described technological innovation as a social process. In our view, what is usually described as 'new media' are not to be regarded as the *agents* of change, but rather as the *outcome* of evolutionary and/or innovative processes within a social domain. The properties of digital representations used within

[58]Examples would be new scientific communication forums (SCF's) in the sense described by Kling et al., (2003). See also chapter 3.

a social domain (e.g. used in the communicative practice of a scientific community) are, in practical terms, determined by these change processes. So it is not the case that the digital medium has certain properties that will inevitably be conferred on any genre that uses it. Rather we shall (and do) see a wide range of different applications where each genre or practice of communication is made to adopt, at any point in time, a specific set of digital properties or 'digitality'. Some practices of communication may indeed be transformed (though not by new media as such, but by choices made by social actors that 'construct' what is perceived as new media), others may stay very close to their traditional modes of representation. This view is supported by the existence of a certain amount of communicative heterogeneity between different scientific domains:[59] there is no single 'digital medium' in science but a whole range of different manifestations of digital properties. None of these has, as we have seen, transformed formal scientific communication insofar as it is based on the peer-reviewed research article. In evolutionary terms, scientific communication has adapted itself by a process of *encapsulation*: through digitization of the journal as a container, the scientific article is able to remain relatively stable even within a digital environment.

Transformation or not, the outcome is not determined (although perhaps restricted) by the properties of the new medium, but by the properties of the specific practice of communication. It therefore makes more sense to adopt a view of the digital format as a 'neutral' concept that can be defined as the availability of certain options to the ongoing evolution of communicative practices. Practitioners may adopt none, some or all of these options to any extent according to the nature and needs of the communicative practice. However, by doing so the communication mechanisms used do not become 'digital media'; what changes is that the mode of communication employs certain digital options. In this sense, the digital medium is hardly a useful concept other than in terms of a technical distribution channel: a digital journal is not *per se* a 'new medium' (in terms of an artifact with a distinct cultural significance), but just another way of distributing scientific articles in the context of formal scientific communication. Or to put it differently: what is often described as the digital medium exists only as the manifestation of digitality in communication practices and their informational genres. In this sense, *the* digital medium does not exist.

As already mentioned, digital information technologies and the specific properties of 'the digital' allow the creation of new practices of communication in the domain of science that require (and could not exist without) these digital means. It is even conceivable that at some point in time these new practices

[59]Kling and McKim 2000; Hyland 2000.

will take over from and replace formal scientific communication by means of the peer-reviewed research article. However, as long as the scientific article exists and performs its key role in scientific communication, its digital manifestation does not necessarily constitute a 'new medium'. And even in the case of a new, digital practice of communication we would argue that it is the scientific community that creates the 'new medium' rather than that the new medium remediates and transforms scientific communication.

3.3 The shadow of the format

One explanation of the limited impact of digitization on the genre of the scientific article may be found in what Nentwich has described as 'the shadow of the format'.[60] The argument here is that the properties of the digital medium influence not only the way a scientist gives expression to facts, ideas, arguments and findings in writing the text, but also influence the substance of research itself, i.e. the choice of research topics, of methodology, data processing etc.[61] Although Nentwich uses this argument to suggest the plausibility that digitization does (or at least will) have an impact on the substance of research, our findings suggest an opposite line of reasoning as far as the actual practice of publishing research results is concerned. If the digital format does indeed tend to exert this type of influence on the way in which scientists choose their subjects, perform their research and create their research writings, this may be felt – however unconsciously – as an unacceptable intrusion of technology on the integrity of science itself, creating a level of resistance that might reinforce rather than diminish the use of traditional forms even in the context of the digital medium.

However, the idea of the 'the shadow of the format' itself is questionable. It is based on a cognitivist view of science that postulates a direct link between the practice of science and the documents it produces. This view regards science as an information processing activity, and the scientific article as the output of that process. In other words: research is seen as an activity aimed at and necessarily leading to specific information products. Within this view, it is natural to think that pre-defined properties of the end-product define the nature of the preceding processes.

In terms of our 3-phase model described in chapter 3, this would mean that the specifications of the output-phase govern the specifications of the input and throughput phases. But the scientific article, although it is indeed an output product of the research process, does not flow 'automatically' from the throughput phase. The scientific article is a *re-writing* of the research process

[60]Nentwich 2003, p. 453–456.

[61]This argument echoes Derrida's idea that 'archivization produces as much as it records the event' ('L'archivation produit autant qu'elle enregistre l'élement' - Derrida 1995, quoted in Ketelaar 1998).

and its results, with specific purposes in mind such as informing, registering, acquiring status and recognition, etc. Most important in this context is the article's role as what Frohmann (1999) has described as 'an objectifying discursive resource'. It is this role that offers a more plausible explanation of the stability of the scientific article as a genre under conditions of digitization.

The article is not a 'literary' form that allows for experimentation and an individualized expression in terms of style, presentation and argumentation. It objectifies by abstracting both from the contingencies of research practice and from the self-expression of the author. As Frohmann writes:

> '[S]cience's literary technology creates resources for the articulation of objectivity, nature, scientific truth, and scientific knowledge. Formal writing is crucial to establishing the documentary techniques for the institutionally authorized enunciation of scientific truth. Studies of scientific practices therefore imply that the journal article is central to such practices, not because it conveys information but because of the centrality of objectifying resources to the cultural phenomenon we know as natural science.'[62]

3.4 The epistemological position

The objectifying role of the scientific article allows us to explain the stability of the scientific article in the light of digitization in more epistemological terms. The peer reviewed article serves in scientific communication as a mechanism that transforms potentially subjective perceptions of the author into certified, objective information in the form of 'justified beliefs'. It can therefore be said that the scientific article 'creates' objective, scientific knowledge. Historically, peer review has been instrumental for the transition from 'subjective' to 'objective' knowledge. As Gross et al., (2002) show, the objectivity of scientific knowledge is not only represented by the process of certification (peer review), but also by structural and stylistic characteristics of the scientific article as a genre. Finally, scientific publishing adds a high degree of persistence to scientific knowledge, both because scientific findings are inscribed on a stable medium, and in a more abstract sense through the 'canonical archive' of interrelated journal articles.[63] This is important because persistence is a requisite for the 'criticizability' of scientific arguments upon which, as Popper notes,[64] the objectivity of scientific knowledge depends.

We have seen in previous chapters that digital formats have a tendency to move in the direction of subjectivity and informality rather than towards objectivity and formality. Not only does the World Wide Web incorporate a vast array of informal communication practices ranging from the popular, trivial, commercial and political to more serious but nevertheless informal practices such as self-publishing by scientists. It also presents itself as a format that

[62]Frohmann 1999, p. 72.

[63]See section 6 on page 63.

[64]Popper 1972, p. 82, 136-137.

tends towards the ephemeral and subjective rather than the persistent and objective. Examples of these properties that we have identified in Chapter 4 include author and reader control over content and presentation, dynamic content, adaptability and functionality. We have also seen that long-term archiving of digital information is problematic. These properties can be perceived as incompatible with the objectifying properties of the scientific article, properties that are not merely 'traditional' and culturally grounded, but also essential in the context of modern science.

The relationship between the process, structure and style of formal scientific communication on the one hand, and scientific objectivity on the other, therefore not only helps to explain the reluctance of scientists to depart from traditional cultural practices. It also points to the fact that new communication practices based on properties of the digital format must, in order to be successful, create acceptable solutions both for the problem of certification and for the problem of finding new structural and stylistics ways of representing and maintaining objectivity.

4. The impact of digitization on scientific communication

4.1 The illusion of a revolution

Why is it that many authors expect digitization to result in a transformation if not a 'revolution' in scientific communication? First, they fail to take into account the social constructivist and evolutionary nature of technological innovation in scientific communication. Their view is based on an autonomous digital medium that will ultimately and inevitably have to be exploited. Second, they also fail to distinguish between the communication process and the substance of scientific information, between the journal as a technical artifact and the article as a genre for representing scientific knowledge. They take for granted that what applies at the aggregate level, will also apply at the level of the individual article and its author. Third, they make the common error of taking second order innovation (revolution) for granted, where first order innovation (evolution) is the rule. Together with views based on social engineering and technological determinism, this leads to representations of the scientific information chain that are prescriptive rather than descriptive, and that tend to advocate unrealistic policies for re-engineering scientific communication.[65] Finally, as we have argued in chapter 2, many authors are politically motivated, seeking 'liberation' of the author and the scientific community from the commercial interests of publishers and other actors in the private sector.

[65] See for instance the recent proposal by Van de Sompel et al., (2004) discussed in chapter 3.

4.2 The dynamics of change

Our own conclusions with respect to the role of technology in the development of scientific communication are summarized in fig. 6.4. Information technology in itself is not a new medium for scientific communication. Rather, it makes available a set of properties – based on enabling technologies – that may or may not be adopted by the academic community and thereby transform scientific communication. The extent to which this happens is governed by two sets of social factors, one constituting pressure towards change, the other constituting resistance to change. In the case of scientific communication, pressure towards change comes from change agents (e.g. key scientists or institutions setting examples of innovation), funding bodies (e.g. putting a premium on more efficient modes of communication), and intermediary actors (e.g. digital libraries, open access publishers). Resistance to change is grounded in the specific culture of scientific disciplines, as well as in existing research and rhetorical practices as described in chapter 2. As Kling et al., (2003, p. 47-48) note, many electronic scholarly communication forums 'have

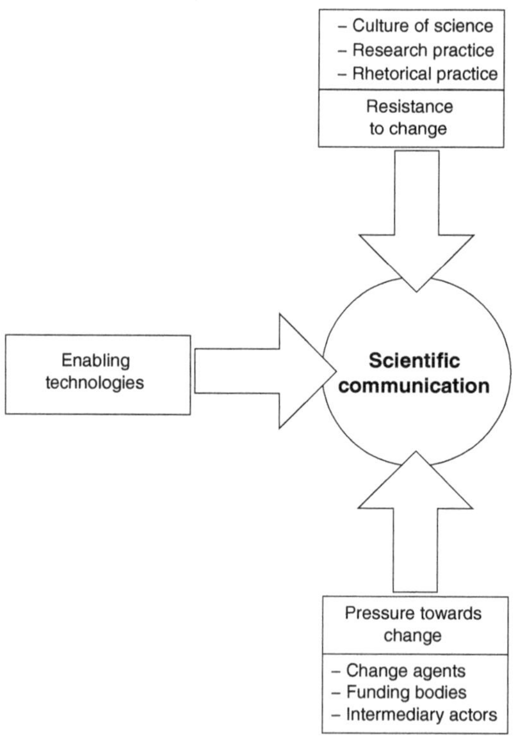

Figure 6.4. Dynamics of change

social properties that are not compatible with highly entrenched and durable scholarly communication practices'.

In this study we have looked primarily at the group of early 'e-only' journals. They can be described as the creation of early adopters of digitization in the domain of scientific communication in the sense that they are precursors to a second stage where digitization has been applied to existing printed journals on a large, and by now in some domains almost complete scale. In terms of the innovation theory described in chapter 2, the creators of these e-only journals might have acted as a change agents, leading the way towards the 'revolution' in scientific communication that many observers seem to have expected. However, as we have seen, that revolution has not happened in terms of a transformation of the scientific article itself, but only in terms of distribution and access. We conclude therefore that as far as the scientific article is concerned, pressures towards digitization have not succeeded in overcoming the resistance to change entrenched in the culture and praxis of science.

5. Final conclusions

The early e-only journal provided a unique forum for experimenting with new forms of formal scientific communication, slightly outside of the mainstream of the existing publishing domain. Many of these journals and their editors earnestly tried to introduce new forms of presentation and interaction made possible by the digital format that might have transformed the journal as a communication forum as well as the genre of the scientific article. However, they did not succeed in setting a convincing example on which the mainstream of scientific publishing could build. Many journals failed to benefit from the possibilities of the digital format, and when they did, authors and readers in general were reluctant to utilize them, even though they used digital means for creating and acquiring scientific information. The e-only journal as an innovative format has now been largely replaced by the even more conventional format of the open access journal. The experiment, in so far as it really was aimed at transforming the scientific article, has failed.

Perhaps the most balanced view is that of Eason et al., (1997, section 6) who, in a relatively early study of the use of e-journals in various disciplines, described the benefits primarily in terms of speed and convenience:

> 'There is broad agreement that being able to search for and retrieve electronic journals from one's place of work is or would be valuable provided needs for personal utility and ease of use are met.
>
> ... Electronic journals are seen by their potential users as one possible service and two kinds of requirement arise. First there is a need in many disciplines for hypertext links not only between journal articles but out of the journals to other information services ... Second, if such links are made, it may be preferable in some disciplines to keep multi-media out of journal articles and locate it in associated databases etc. This makes it possible to preserve the form, quality, length etc. of traditional journals which is a desire in many disciplines.
>
> ... For most of the text based disciplines and some of the others the most desirable form of electronic journal delivery system is one in which a paper can be found, its full text can be delivered

> electronically and it can be printed to create a personal copy ... This is the current level of expectation of many scholars which may be changed as more evidence emerges of other possible services of value to the discipline.

Claims that 'the scientific journal will change considerably [towards a] new, more composite form as an ensemble of various textual and non-textual components'[66] or that the scientific article will develop towards a new hypertextual form[67] cannot be substantiated on the basis of this study. It might be argued that the lack of innovation of the scientific journal in digital form can be ascribed to a declining importance of the genre as such. But there are no indications that the role of the scientific article is diminishing, and the recent growth in numbers of new, open access journals serves to prove the continuing attractiveness of the genre. As Van Raan notes:

> 'Electronic publishing developments and new information technology in general will affect the main functions of scientific communication. Most changes however will be primarily technological but not conceptual. Publication via journals of high reputation is in most fields of science crucial to receive professional recognition. That will remain so in the 'electronic era'.'[68]

In our innovation study that coincided with the publication of the first electronic journals, we concluded that until then the application of ICT had impacted on the *distribution channel* of scientific information, but meant relatively little for the genre of the article itself: 'at the end of the process, [the end-user] will still obtain a copy of an article from a published journal'.[69] Our findings in this study confirm that little has changed in this respect since the wide-scale adoption of information technology in scientific communication. This leads us to the interesting conclusion that scientists do accept innovative methods as far as searching and acquiring information is concerned, but far less so when creating or processing information.

The scientific journal remains a key genre in the dissemination of scientific knowledge, and the impact of digitization in this field is therefore an issue of some importance. The scientific information chain is currently involved in a process of innovation, especially at the level of the journal and at the aggregate level of the information systems created by publishers, content providers and libraries. It is here that we see digitization giving rise to new modes of distribution, new business models and new user-oriented functions. These value-adding innovations have been adopted by traditional as well as new commercial publishers and scholarly societies. However, there is no evidence that digitization has transformed the practice of scientific authorship at the level of the journal article. This study has found that authors show little interest in adopting new ways of documenting and presenting research results by means of the

[66] Kircz 2001.
[67] Nentwich 2003.
[68] Raan 2001.
[69] Mackenzie Owen and van Halm 1989, p. 76.

research article, or in deviating too far from the traditional culture of scientific communication. In the same way as the invention of the printing press introduced a new technology of reproduction with initially little impact on the content of what was reproduced, digitization functions as a new technology of distribution, again with little impact on the content of what is distributed.

Therefore, we conclude that the impact of digitization on formal scientific communication is to be found in important enhancements of the communication system, rather than in the substance of scientific information itself. The key actors in this process of innovation are the intermediary actors such as publishers and libraries,[70] rather than the community of scientific authors. The main beneficiaries of digitization are the end-users of the communication system through improved access to a key genre of scientific information that has remained relatively unchanged in the age of digitization.

[70]with publishers as the main change agent, in line with our socio-technical network analysis of innovation described in 9.2.

Bibliography

Abramson, A. (2000). Solutions for delivering digital content in the new academic enterprise. *Library hi-tech news* 17(16–18):1.

Aitchison, T. M. (1988). The database producer in the information chain. *Journal of information science* 14(6):319–327.

Allen, Colin, Uri Nodelman, and Edward N. Zalta (2002). The Stanford Encyclopedia of Philosophy: a developed dynamic reference work. *Metaphilosophy* 33(1-2):210–228, reprinted in CyberPhilosophy: The Intersection of Philosophy and Computing, James H. Moor and Terrell Ward Bynum, (eds.), Oxford: Blackwell, pp. 201-218.

Atherton, J.S. (2003). Learning and teaching: cognitive dissonance. http://www.dmu.ac.uk/~jamesa/learning/dissonance.htm

Atkinson, Ross (2000). A rationale for the redesign of scholarly information exchange. *Library resources & technical services* 44:59–69.

AUCC (1995). Towards a new paradigm for scholarly communication: discussion paper prepared by the AUCC - CARL/ABRC Task Force on Academic Libraries and Scholarly Communication. http://library.uwaterloo.ca/documents/scholarly(aucc-carl).html

Bachrach, Steven M. (2001). Scientific journals of the future. In *The transition from paper: where are we going and how will we get there*, R. Stephen Berry and Anne Simon Moffat, eds., American Academy of Arts & Sciences. http://www.amacad.org/publications/trans4.htm

Bartunek, J.M. and M.K. Moch (1987). First-order, second-order, and third-order change and organizational development interventions: a cognitive approach. *Journal of Applied Behavioral Science* 23:483–500.

Bates, Marcia J. (1999). The invisible substrate of information science. *Journal of the American Society for Information Science* 50:1043–1050. http://www.gseis.ucla.edu/faculty/bates/substrate.html

Baudrillard, Jean (1983). *Simulations*. New York: Semiotext(e).

Bazerman, Charles (1988). *Shaping written knowledge: the genre and activity of the experimental article in science*. University of Wisconsin Press. http://wac.colostate.edu/books/bazerman_shaping/

Bearman, D. (1999). Reality and chimeras in the preservation of electronic records. *D-Lib magazine* 5. http://www.dlib.org/dlib/april99/bearman/04bearman.html

Becker, E., W. Buhse, D. Günnewig, and N. Rump, eds. (2003). *Digital rights management: technological, economic, and legal and political aspects*. No. 2770 in Lecture notes in computer science, Berlin: Springer.

Belkin, N.J. (1990). The cognitive viewpoint in information science. *Journal of information science* 16:11–15.

Benedek, J (1970). Scientific-political principles, the scientific-technical revolution and its contact with technical libraries. *Konyvtari figyelo* 16(4):292–301.

Berners-Lee, Tim, James Hendler, and Ora Lassila (2001). The Semantic Web: A new form of Web content that is meaningful to computers will unleash a revolution of new possibilities. *Scientific American* (May).

Bethesda (2003). Bethesda Statement on Open Access Publishing. http://www.earlham.edu/~peters/fos/bethesda.htm

Bijker, Wiebe E. (1995). *Of bicycles, bakelites and bulbs: towards a theory of sociotechnical change*. MIT Press.

Bijker, Wiebe E. (2001). Social construction of technology. In *International encyclopedia of the social & behavioral sciences*, vol. 23, pp. 15522–15527.

Bijker, Wiebe E., Thomas. P. Hughes, and Trevor Pinch, eds. (1987). *The social construction of technological systems: new directions in the sociology and history of technology*. MIT Press.

Billing, Christian (2004). Modelling the anatomy theatre and the indoor hall theatre: dissection on the stages of early modern London. *Early modern literary studies* Special issue 13:3.1–17. http://purl.oclc.org/emls/si-13/billing

Birkerts, S. (1994). *The Gutenberg elegies: The fate of reading in an electronic age*. Boston, MA: Faber & Faber.

Bishop, Ann Peterson (1998). Digital libraries and knowledge disaggregation: the use of journal article components. In *DL98: proceedings of the 3rd ACM International Conference on Digital Libraries*, New York: ACM. http://dli.grainger.uiuc.edu/dlisoc/socsci_site/conf-dl98-ann-knowl-disag.html

Björk, Bo-Christer (2004). Open access to scientific publications - an analysis of the barriers to change. *Information Research* 9(2). http://informationr.net/ir/9-2/paper170.html

Björk, Bo-Christer and Turid Hedlund (2004). A formalised model of the scientific publication process. *Online information review* 28(1):8–21.

Björk, Bo-Christer, Turid Hedlund, and Tomas Gustafsson (2002). SciX: scientific publishing: as-is business and information model, sciX Deliverable D1. http://www.scix.net/d/01/d1.pdf

Bolter, Jay David (1991). *Writing space: the computer, hypertext, and the history of writing*. Hillsdale, NJ: Erlbaum.

Bolter, Jay David (2001). *Writing space: computers, hypertext, and the remediation of print*. Mahwah, NJ: Erlbaum.

Bolter, Jay David and Richard Grusin (1999). *Remediation: understanding new media*. MIT Press.

Bonnevie, Ellen (2001). Dretske's semantic information theory and metatheories in library and information science. *Journal of documentation* 57(4):519–534.

Borko, Harold (1968). Information science - what is it? *Journal of the American Society for Information Science* 19(1):3–5.

Bowker, Geoffrey C and Susan Leigh Star (1999). *Sorting things out: classification and its consequences*. Cambridge: MIT Press.

Bra, Paul de, Peter Brusilovsky, and Geert-Jan Houben (1999). Adaptive hypermedia: from systems to framework. *ACM Computing Surveys* 31(4). http://www.cs.brown.edu/memex/ACM_HypertextTestbed/papers/25.html

Branin, J.J. and M. Case (1998). Reforming scholarly publishing in the sciences: a librarian perspective. *Notices of the AMS*. http://www.ams.org/notices/199804/branin.pdf

Briggs, Asa and Peter Burke (2002). *A social history of the media: from Gutenberg to the Internet.* Polity Press.

Brookes, B.C. (1980). The foundations of information science, part I: philosophical issues. *Journal of information science* 2:125–133.

Brown, Cecelia (2003). The changing face of scientific discourse: analysis of genomic and proteomic database usage and acceptance. *Journal of the American Society for Information Science and Technology* 54(10):926–938.

Brown, Harcourt (1967). *Scientific organizations in seventeenth century France.* New York.

Brown, John Seeley and Paul Duguid (2000). *The social life of information.* Boston: Harvard Business School Press.

Bruland, Kristine (2001). Technological revolutions, innovation systems and convergence from a historical perspective. Department of History, University of Oslo. http://pascal.iseg.utl.pt/{\textasciitilde}converge/pdfs/(45).pdf

Brusilovsky, Peter (1996). Methods and techniques of adaptive hypermedia. *User modeling and user-adapted interaction* 6(2-3):87–129. http://www2.sis.pitt.edu/eterb/papers/UMUAI96.pdf

Brusilovsky, Peter, Alfred Kobsa, and Julita Vassileva (1998). *Adaptive hypertext and hypermedia.* Dordrecht: Kluwer Academic Publishers.

Buck, A.M., R.C. Flagan, and B. Coles (1999). Scholars forum: a new model for scholarly communication. http://library.caltech.edu/publications/scholarsforum/

Buckland, Michael K. (1991a). *Information and information systems.* Westport: Greenwood.

Buckland, Michael K. (1991b). Information as thing. *Journal of the American Society for Information Science* 42(5):351–360.

Buckland, Michael K. (1998). What is a digital document? *Document numérique* 2(2), preprint. http://www.sims.berkeley.edu/~Buckland/digdoc.html

Budd, John M. (2001). *Knowledge and knowing in library and information science: a philosophical framework.* Lanham, MD: Scarecrow.

Burg, Jennifer, Yue-Ling Wong, Ching-Wan Yip, and Anne Boyle (2000). The state of the art in interactive multimedia journals for academia. *IMEJ of*

Computer-Enhanced Learning ED-MEDIA 2000, AACE, June 2000, Montreal, Quebec, Canada.
http://imej.wfu.edu/articles/EDMEDIA2000paper/index.asp

Burke, Peter (2000). *A social history of knowledge: from Gutenberg to Diderot.* Polity Press.

Campbell, Donald T. (1956). Perception as substitute trial and error. *Psychological review* 63(5):331–342.

CEDARS Project (2002a). Cedars Guide to Digital Collection Management.
http://www.leeds.ac.uk/cedars/guideto/collmanagement/guidetocolman.pdf

CEDARS Project (2002b). Cedars Guide to Digital Preservation Strategies.
http://www.leeds.ac.uk/cedars/guideto/dpstrategies/dpstrategies.html

Charles, Peter, Nathan Good, Laheem Lahmar Jordan, and Joyojeet Pal (2003). How much information 2003.
http://www.sims.berkeley.edu/research/projects/how-much-info-2003/printable_report.pdf

Chatman, Elfredda A. (1996). The impoverished life-world of outsiders. *Journal of the American Society for Information Science* 47(3):193–206.

Chien, Y.T. (1997). Digital libraries, knowledge networks, and human-centered information systems.
http://www.dl.ulis.ac.jp/ISDL97/proceedings/ytchien/ytchien.html

Chodorow, Stanley (1998). The faculty, the university, and intellectual property. *Journal of electronic publishing* 3.
http://www.press.umich.edu/jep/03-03/chodorow.html

Cisne, John L. (2005). How science survived: medieval manuscripts, demography and classic texts extinction. *Science* 307(5713):1305–1307.
http://www.sciencemag.org/cgi/content/abstract/307/5713/1305

Colby, Charles W., ed. (1920). *Selections from the sources of English history.* London: Longmans, Green & Co.

Consortium for Educational Technology for University Systems (1997). *The academic library in the information age: changing roles.* California State University. http://www.cetus.org/acad_lib.pdf

Cooke, Helen (2004). A historical review of the chemistry periodical literature until 1950. *Learned Publishing* 17(2):125–134.

Corning, Mary E. and Martin M. Cummings (1976). Biomedical communications. In *Advances in American medicine*, vol. 2, John Z. Bowers and Elizabeth F. Purcell, eds., New York: Josiah Macy.

Cox, John E. (1999). *Publisher/library relationships in the digital environment.* STM International Association of Scientific, Technical and Medical Publishers.

Crawford, Walt (2002). Free electronic refereed journals: getting past the arc of enthousiasm. *Learned publishing* 15:117–226.

Crow, Raym (2002). *The case for institutional repositories: a SPARC position paper.* http://www.arl.org/sparc/IR/IR_Final_release_102.pdf

Crowther, A. (1999). Consortia licensing, information as infrastructure. *IATUL proceedings (new series). 8 1999, CD-ROM Full Text Database.*

Cummings, A.M. et al., (1992). University libraries & scholarly communication: a study prepared for the Andrew W. Mellon Foundation. ARL. http://www.arl.org/scomm/mellon

Day, Ron (1997). Paul Otlet´s book and the writing of social space. *Journal of the American Society for Information Science* 48(4):310–317.

Day, Ronald E. (2001). *The modern invention of information: discourse, history, and power.* Southern Illinois University Press.

Derrida, J. (1995). *Mal d'archive: une impression freudienne.* Paris, Galilee, translated as 'Archive fever' (Chicago, London, 1996).

Dicks, Bella and Bruce Mason (1998). Hypermedia and ethnography: reflections on the construction of a research approach. *Sociological research online* 3(3). http://www.socresonline.org.uk/socresonline/3/3/3.html

Diessen, R.J. van and T. van der Werf-Davelaar (2002). *Authenticity in a digital environment.* http://www.kb.nl/kb/ict/dea/ltp/reports/2-authenticity.pdf

Dillon, Andrew, Cliff McKnight, and John Richardson (1990). Navigation in hypertext: a critical review of the concept. In *INTERACT '90 – Proceedings of the IFIP TC13 Third International Conference on Human-Computer Interaction*, pp. 587–592, North-Holland.

Dovey, Matthew (1999). Meta-objects: an object-oriented approach to metadata. *Ariadne* (19). http://www.ariadne.ac.uk/issue19/meta-objects/

Dowling, W.C. (1997). Saving scholarly publishing in the age of Oprah: The Glastonbury project. *Journal of scholarly publishing* 28:115.

Dretske, F.I. (1981). *Knowledge and the flow of information*. Oxford: Blackwell.

Duff, A.S. (1997). Some post-war models of the information chain. *Journal of librarianship and information science* 29(4):179–187.

Duff, A.S. (1998). Daniel Bell's theory of the information society. *Journal of information science* 24:373–393.

Duguid, Paul (1996). Material matters: aspects of the past and the futurology of the book. In *The future of the book*, Geoffrey Nunberg, ed., University of California Press. http://www.slofi.com/Material_Matters.htm

Eamon, William (1994). *Science and the secrets of nature: books of secrets in medieval and early modern culture*. Princeton University Press.

Eason, Ken, Chris Carter, Susan Harker, Sue Pomfrett, Kathy Phillips, and John Richardson (1997). *A comparative analysis of the role of multi-media electronic journals in scholarly disciplines*. HUSAT Research Institute and Department of Human Sciences, Loughborough University. http://www.ukoln.ac.uk/services/elib/papers/supporting/pdf/eason.pdf

Education for Change (2002). Researcher use of libraries and other information sources: current patterns and future trends. http://www.rslg.ac.uk/research/libuse

Edwards, Deborah M. and Lynda Hardman (1999). Lost in hyperspace: cognitive mapping and navigation in a hypertext environment. In *Hypertext: theory into practice*, Ray McAleese, ed., pp. 90–105, Intellect Books.

van Eemeren, F.H., R. Grootendorst, Sally Jackson, and Scott Jacobs (1993). *Reconstructing argumentative discourse: studies in rhetoric and communication*. University of Alabama Press.

Eisend, M (2002). The Internet as a new medium for the sciences? The effects of Internet use on traditional scientific communication media among social scientists in Germany. *Online information review* 26(5):307–317.

Eisenstein, Elisabeth L. (1980). *The printing press as an agent of change: communications and cultural transformation in early-modern Europe, volumes I and II*. Cambridge University Press, 1st paperback ed., first published in two volumes 1979.

Electronic Privacy Information Center (2004). Digital rights management and privacy. http://www.epic.org/privacy/drm/

Elsevier (1996). TULIP final report.
http://www.elsevier.com/wps/find/librariansinfo.librarians/tulipfr

Faraj, S. and M. M. Wasko (2001). The web of knowledge: An investigation of knowledge exchange in networks of practice.
http://opensource.mit.edu/papers/Farajwasko.pdf

Federman, Mark (2003). The cultural paradox of the global village. In *Panel on Digitization of Information and the Future of Culture at the EU-Japan Fest 10th Anniversary Symposium on The Role of Culture in an Age of Advancing Globalization, held in Tokyo, Japan, February 10-11, 2003.*
http://www.mcluhan.utoronto.ca/article_culturalparadox.htm

Ferrero, Giovanni (2001). Preface. In *Say not to say: new perspectives on miscommunication*, L. Anolli, R. Ciceri, and G. Riva, eds., Amsterdam: IOS Press.

Ferris, Sharmila Pixy (2002). Writing electronically: the effects of computers on traditional writing. *Journal of electronic publishing* 8.
http://www.socresonline.org.uk/3/3/3.html

Festinger, Leon (1957). *A theory of cognitive dissonance*. Evanston: Row Peterson.

Fjällbrant, Nancy (1997). Scholarly communication: historical development and new possibilities.
http://internet.unib.ktu.lt/physics/TEXTS/schoolarly/scolcom.htm

Fog, Agner (1999). *Cultural selection*. Kluwer Academic Publishers.
http://www.agner.org/cultsel/

Foroughi, A., M. Albin, and S. Gillard (2002). Digital rights manangement: a delicate balance between protection and accessibility. *Journal of information science* 28(5):380–395.

Foskett, Douglas J. (1978). The theory of integrative levels and its relevance to the design of information systems. *Aslib Proceedings* 30(6):202–208.

Foucault, Michel (2002). *The archaeology of knowledge*. London and New York: Routledge Classics, first published Paris: Gallimard, 1969. Translated by A.M. Sheridan Smith.

Franstvag, Jan Erik (2002). Re-forging the value chain in scientific publishing.

Freeman, J.B. (1991). *Dialectics and the macrostructure of arguments; a theory of argument structure*. Berlin: Foris.

Friedlander, Amy and Rändi S. Bessette (2003). *The implications of information technology for scientific journal publishing: a literature review*. National Science Foundation, Division of Science Resources Statistics. http://www.nsf.gov/sbe/srs/nsf03323/

Friend, F.J. (1998). Alternatives to commercial publishing for scholarly communication. *Serials* 11(2):163–166.

Frohmann, Bernd (1999). The role of the scientific paper in science information systems. In *History of information science: proceedings of the 1998 Conference on the History and Heritage of Science Information Systems*, Mary Ellen Bowden, Trudi Bellardo Hahn, and Robert V. Williams, eds., Medford, NJ: Information Today. http://www.chemheritage.org/explore/ASIS_documents/ASIS98_Frohmann.pdf

Füssel, Stephan (2001). Gutenberg and today's media change. *Publishing research quarterly* 16(4):3–10.

Garvey, W.D. (1979). *Communication: the essence of science*. Oxford: Pergamon Press.

Garvey, W.D. and B.C. Griffith (1972). Communication and information processing within scientific disciplines: empirical findings for psychology. *Information storage and retrieval* 8:123–126.

Gass, S. (2001). Transforming scientific communication for the 21st century. *Science & technology libraries* 19:3–18.

Gervais, Daniel J. (1999). Electronic rights management and digital identifier systems. *Journal of electronic publishing* 4(3). http://www.press.umich.edu/jep/04-03/gervais.html

Giles, M.W. (1996). From Gutenberg to gigabytes: scholarly communication in the age of cyberspace. *Journal of politics* 58:613.

Ginsparg, Paul (2001). Creating a global knowledge network. In *Second Joint ICSU Press - UNESCO Expert Conference on Electronic Publishing in Science, Paris, 19-23 Feb 2001*. http://arxiv.org/blurb/pg01unesco.html

Glaser, B.G. and A.L. Strauss, eds. (1967). *The discovery of grounded theory: strategies for qualitative research*. Chicago: Aldine Publishing.

Glazier, Jack D. and Robert Grover (2002). A multidisciplinary framework for theory building. *Library trends* 50(3):317–329.

Gnoli, Claudio (2003). Taxology: classification by naturalistic approach. (See section 32: Integrative levels). http://www-dimat.unipv.it/~gnoli/taxology/

Gorniak-Kocikowska, K (2001). Revolution and the library. *Library trends* 49(3):454–470.

Gould, Stephen Jay (2004). *The hedgehog, the fox and the master´s pox: mending and minding the misconceived gap between science and the humanities.* Vintage, first published by Jonathan Cape, 2003.

Granger, Stewart (2000). Emulation as a Digital Preservation Strategy. *D-lib magazine* 6. http://www.dlib.org/dlib/october00/granger/10granger.html

Granger, Stewart (2002). Digital preservation & emulation: from theory to practice. ICHIM conference (http://www.archimuse.com/ICHIM01) Milan, Sept 3-7. http://www.leeds.ac.uk/cedars/pubconf/papers/ichim01SG.html

Greene, Sarah and Matthew Cockerill (1997). The BioMedNet and HMS Beagle Models. In *First International Virtual Conference on Infectious Diseases of Animals*, National Animal Disease Center, Ames, IA, USA. http://www.nadc.ars.usda.gov/virtconf/keynote/keynote4.htm

Gross, Alan (1990). *The rhetoric of science.* Harvard University Press.

Gross, Alan G., Joseph E. Harmon, and Michael Reidy (2002). *Communicating science: the scientific article from the 17th century to the present.* Oxford University Press.

Gunn, Moira A. (2000). *The societal consequences of digitalization.* Working Paper of the Research Group on the Global Future, Center for Applied Policy Research. http://www.cap-info.de/triangle/download/digworld.PDF

Guédon, Jean-Claude (2001). *In Oldenburg's long shadow: librarians, research scientists, publishers, and the control of scientific publishing.* Washington: ARL. http://www.arl.org/arl/proceedings/138/guedon.html

Hall, Stuart (1980). Encoding/decoding. In *Culture, media, language: working papers in cultural studies, 1972-1979*, Centre for Contemporary Cultural Studies, ed., London: Hutchinson, originally published in 1973.

Halliday, Leah and Charles Oppenheim (2001). Developments in digital journals. *Journal of documentation* 57:260–283.

Halporn, B. (1997). The scholarly monograph in crisis. *College and research libraries news* 58(10):706–7.

Hammes, M. (2001). Beyond e: scholarly communication in the knowledge era. *Mousaion* 19(2):45–60.

Harmon, Joseph E. and Alan G. Gross (2003). *The scientific article: from Galileo's New Science to the human genome*. The Fathom Archive. http://www.fathom.com/course/21701730/index.html

Harmsze, F.A.P., M.C. van der Tol, and J.G. Kircz (1999). A modular structure for electronic scientific articles. In *Conferentie Informatiewetenschap 1999. Centrum voor Wiskunde en Informatica, Amsterdam, 12 november 1999*, no. 99-20 in Computing Science Reports, pp. 2–9, Dept. of Mathematics and Computing Science. Technische Universiteit Eindhoven. http://wwwis.win.tue.nl/infwet99/proceedings/harmsze.html

Harmsze, Frédérique (2000). *A modular structure for scientific articles in an electronic environment*. Ph.D. thesis, Universiteit van Amsterdam. http://www.science.uva.nl/projects/commphys/papers/thesisfh/Front.html

Harnad, Stevan (1991). Post-Gutenberg galaxy: the fourth revolution in the means of production of knowledge. *Public-access computer systems review* 2(1):39–53. http://cogprints.ecs.soton.ac.uk/archive/00001580/00/harnad91.postgutenberg.html

Harnad, Stevan (1992). The Turing Test is not a trick: Turing indistinguishability is a scientific criterion. *SIGART Bulletin* 3(4):9–10. http://www.ecs.soton.ac.uk/arnad/Papers/Harnad/harnad92.turing.html

Harnad, Stevan (1996). Implementing peer review on the net: scientific quality control in scholarly electronic journals. pp. 103–108. http://cogprints.ecs.soton.ac.uk/archive/00001692/00/harnad96.peer.review.html

Hartley, Harold, ed. (1960). *The Royal Society: its origins and founders*. London: The Royal Society.

Hatch, Robert A. (2002). The scientific paradigm - paradigm lost?, (An earlier and shorter version of this brief historiographic essay was published in the OAH Magazine of History, 4, 2 (1989): 34-39). http://web.clas.ufl.edu/users/rhatch/pages/03-Sci-Rev/SCI-REV-Home/08sr-htch.htm

Hauben, Michael and Ronda Hauben (1997). *Netizens: on the history and impact of Usenet and the Internet*. Wiley-IEEE Computer Society Press.

Hedlund, Turid, Tomas Gustafsson, and Bo-Christer Björk (2004). The open access scientific journal: an empirical study. *Learned publishing* 17(3):199–209.

Henderson, Albert (2002a). Diversity and the growth of serious/scholarly scientific journals. In *Scholarly publishing: books, journals, publishers and libraries in the twentieth century*, Richard E Abel and Lyman W. Newlin, eds., pp. 133–161, Wiley.

Henderson, Albert (2002b). The growth of printed literature in the twentieth century. In *Scholarly publishing: books, journals, publishers and libraries in the twentieth century*, Richard E Abel and Lyman W. Newlin, eds., pp. 1–23, Wiley.

Henshaw, Robin (2001). What next for internet journals? Implications of the trend towards paid placement in search engines. *First Monday* 6(1). http://www.firstmonday.dk/issues/issue6_9/henshaw/index.html

Hibbitts, B. (1999). From law reviews to knowledge networks: legal scholarship in the age of cyberspace. *Serials review* 25(1):1–9.

Hitchcock, Steve, Leslie Carr, and Wendy Hall (1996). A survey of STM online journals: the calm before the storm. In *Directory of electronic journals, newsletters and academic discussion lists*, D. Mogge, ed., Washington: ARL, 1996, 6 ed. http://journals.ecs.soton.ac.uk/survey/survey.html

Hjørland, Birger (1998). Theory and metatheory of information science: a new interpretation. *Journal of documentation* 54(5):606–621.

Hjørland, Birger (2002a). Domain analysis in information science - Eleven approaches - traditional as well as innovative. *Journal of documentation* 58(4):422–462.

Hjørland, Birger (2002b). Epistemology and the socio-cognitive perspective in information science. *Journal of the American Society for Information Science and Technology* 53(4):257–270.

Holdsworth, David and Paul Wheatley (2001). Emulation, preservation and abstraction. http://129.11.152.25/CAMiLEON//dh/ep5.html

Hooker, Richard (1996). The European enlightenment: the scientific revolution. http://www.wsu.edu:8080/ee/ENLIGHT/SCIREV.HTM

Houghton, John W. (2001). Crisis and transition: the economics of scholarly communication. *Journal of electronic publishing* 14:167–176.

Houghton, John W., Colin Steele, and Margaret Henty (2003). Changing Research Practices in the Digital Information and Communication Environment. Canberra: Department of Education, Science and Training.

Houghton, John W., Colin Steele, and Margaret Henty (2004). Research practices and scholarly communication in the digital environment. *Learned publishing* 17(3):231–249.

Hubert, Agnès and Bénédicte Caremier (2000). *Democracy and the information society in Europe*. Kogan Page - Office for Official Publications of the European Communities.

Hull, David L. (1990). *An evolutionary account of the social and conceptual development of science*. University of Chicago Press, pbck. ed., first published in 1988.

Hull, David L. (2000). *Essays on biological evolution and the philosophy of science*. Cambridge University Press.

Hummels, H. and H. E. Roosendaal (2001). Trust in scientific publishing. *Journal of business ethics* 34:87–100.

Hunter, Michael (1989). *Establishing the new science: the experience of the early Royal Society*. Woodbridge: Boydell Press.

Hunter, Michael (1994). *The Royal Society and its Fellows, 1660-1700: the morphology of an early scientific institution*. Chalfont St.Giles: British Society for the History of Science, 2nd ed. ed.

Hunter, Philip (2001). The management of content: universities and the electronic publishing revolution. *Ariadne* 28.
http://www.ariadne.ac.uk/issue28/cms/

Huntley, Arthur C. (1995). The skin and diabetes mellitus. *Dermatology online journal* 1:2.
http://dermatology.cdlib.org/DOJvol1num2/diabetes/diabetes-title.html

Hurd, J. M. (2000). The transformation of scientific communication: A model for 2020. *Journal of the American Society for Information Science* 51:1279–1283.

Hyland, Ken (2000). *Disciplinary discourses: social interactions in academic writing*. Pearson Education.

Ingwersen, P. (1996). Cognitive perspectives of information retrieval interaction. *Journal of documentation* 52(1):3–50.

Jakobsen, Roman (1960). Closing statement: linguistics and poetics. In *Style in language*, Thomas A. Sebeok, ed., pp. 350–377, MIT Press.

Jakobsen, Roman and Morris Halle (1956). *Fundamentals of language*. The Hague: Mouton.

Jobson, E. (2003). Digital printing: current and future applications. *Publishing research quarterly* 19(1):20–30.

Johns, Adrian (2000). Miscellaneous methods: authors, societies and journals in early modern England. *British journal for the history of science* 33:159–186. http://www.nd.edu/~dharley/HistIdeas/texts/Johns-authors.pdf

Judge, P.J. (1967). The user-system interface today: national and international information systems. In *Communication in science: documentation and automation*, Anthony De Reuck and Julie Knight, eds., pp. 37–51, London: J.&A. Churchill.

Kaminer, Y and N. Braunstein (1998). Bibliometric analysis of the impact of Internet use on scholarly productivity. *Journal of the American Society for Information Science* 49:720–730.

Katzenbeisser, Stefan and Fabien A.P. Petitcolas, eds. (1999). *Information hiding techniques for steganography and digital watermarking*. Artech House.

Kaufer, D.S. and K.M. Carley (1993). *The influence of print on sociocultural organization and change*. Hillsdale, N.J., LEA.

Kaufman, P. (1998). Structure and crisis: markets and market segmentation in scholarly publishing. In *The mirage of continuity: reconfiguring academic information resources for the 21st century*, pp. 178–192.

Kearney, Hugh F. (1966). *Origins of the scientific revolution*. London.

Keller, Albert Galloway (1915). *Societal evolution: a study of the evolutionary basis of the science of society*. New York, Macmillan.

Keller, Alice (2001). Future development of electronic journals: a Delphi survey. *Electronic library* 19(6):383–396.

Ketelaar, F.C.J. (1998). *Archivalisering en archivering*. Samsom.

Kiernan, V. (1999). Why do some electronic-only journals struggle, while others flourish? *Journal of electronic publishing* 4.
http://www.press.umich.edu/jep/04-04/kiernan.html

Kim, H. J. (2001). The transition from paper to electronic journals: key factors that affect scholars' acceptance of electronic journals. *Serials librarian* 41(1):31–64.

Kircz, Joost G. (1998). Modularity: the next form of scientific information presentation? *Journal of documentation* 54(2):210–235.
http://www.kra.nl/Website/Artikelen/Jdoc98.htm

Kircz, Joost G. (2001). New practices for electronic publishing 1: will the scientific paper keep its form? *Learned publishing* 14:265–272.

Kircz, Joost G. (2002). New practices for electronic publishing 2: new forms of the scientific paper. *Learned publishing* 15(1):27–32.

Kircz, Joost G. and F.A.P. Harmsze (2000). Modular scenarios in the electronic age. In *Proceedings Conferentie informatiewetenschap 2000. Doelen, Rotterdam 5 april 2000*, Paul van der Vet and Paul de Bra, eds., no. 00-20 in CS-Reports, pp. 31–43. http://www.kra.nl/Website/Artikelen/mod2k.html

Kircz, Joost G. and Hans E. Roosendaal (1996). Understanding and shaping scientific information transfer. In *Electronic publishing in science: proceedings of the joint ICSU Press/UNESCO Expert Conference, February 1996*, pp. 106–116. http://www.library.uiuc.edu/icsu/kircz.htm

Kling, Rob (1999). What is Social Informatics and why does it matter? *D-Lib magazine* January. http://www.dlib.org/dlib/january99/kling/01kling.html

Kling, Rob and Ewa Callahan (2003). Electronic journals, the Internet, and scholarly communication. http://www.slis.indiana.edu/csi/WP/wp01-04B.html

Kling, Rob and Lisa Covi (1995). Electronic journals and legitimate media in the systems of scholarly communication. *Information society* 11(4):261–271.

Kling, Rob and Roberta Lamb (1996). Analyzing visons of electronic publishing and digital libraries. http://www.slis.indiana.edu/kling/pubs/EPUB6.htm

Kling, Rob and Geoffrey McKim (1999). Scholarly communication and the continuum of electronic publishing. *Journal of the American Society for Information Science* 50:890–906.

Kling, Rob and Geoffrey McKim (2000). Not just a matter of time: field differences in the shaping of electronic media in supporting scientific communication. *Journal of the American Society for Information Science* 51:1306. http://xxx.lanl.gov/ftp/cs/papers/9909/9909008.pdf

Kling, Rob, Geoffrey McKim, and A. King (2003). A bit more to it: scholarly communication forums as socio-technical interaction networks. *Journal of the American Society for Information Science and Technology* 54:47.

Kling, Rob, Lisa Spector, and Geoff McKim (2002). Locally publishing in the Internet: the GUILD model. *Journal of electronic publishing* 8.

Korpela, Mikko, Anja Mursu, and H. Abimbola Soriyan (2002). Two times four integrative levels of analysis: a framework.
http://www.uku.fi/atkk/actad/2x4.pdf

Kronick, David A. (1976). *A history of scientific and technical periodicals: the origins and development of the scientific and technical press, 1665-1790.* Scarecrow Press, 2 ed., originally published in1962.

Kronick, David A. (1991). Peer review in 18th century scientific journalism. In *Peer review in scientific publishing*, Chicago: Council of Biology Editors.

Kronick, David A. (2001). The commerce of letters: networks and "invisible colleges" in seventeenth- and eighteenth-century Europe. *Library quarterly* 71:28–43.

Kuhn, Thomas S. (1996). *The structure of scientific revolutions.* University of Chicago Press, 3rd ed., originally published in 1962.

LaFollette, Marcel (1992). *Stealing into print: fraud, plagiarism, and misconduct in scientific publishing.* University of California Press.

Lakoff, George and Mark Johnson (1980). *Metaphors we live by.* Chicago: Univ. of Chicago Press.

LaMacchia, Brian A. (2002). Key challenges in DRM: an industry perspective. In *ACM DRM Workshop, Oct 15, 2002.*
http://crypto.stanford.edu/DRM2002/abstract-bal.doc

Lancaster, F.W. (1978). *Towards paperless information systems.* New York: Academic Press.

Lancaster, F.W. (1995). The evolution of electronic publishing. *Library trends* 43(4):518–527.

Landow, G.P. (1992). *Hypertext: the convergence of contemporary critical theory and technology.* Baltimore, MD: Johns Hopkins University Press.

Lanham, R. (1994). *The electronic word: democracy, technology, and the arts.* University of Chicago.

Latour, Bruno (1987). *Science in action: how to follow scientists and engineers through society.* Harvard University Press.

Lederberg, Joshua (1996). Options for the future. *D-Lib Magazine, May 1996* 2(5). http://www.dlib.org/dlib/may96/05lederberg.html

Levy, David M. (1994). Fixed or fluid?: document stability and new media. pp. 24–31. http://doi.acm.org/10.1145/192757.192760

Leydesdorff, Loet (2001). The complex dynamics of scientific communication. In *Language - meaning - social construction: interdisciplinary studies,*

Colin B. Grant and Donal McLaughlin, eds., pp. 163–179, Amsterdam/New York: Rodopi. http://users.fmg.uva.nl/lleydesdorff/scicomm

Liu, Ziming (2003). Trends in transforming scholarly communication and their implications. *Information processing & management* 39(6):889–898.

Llewellyn, Richard D., Lorraine J. Pellack, and Diana D. Shonrock (2002). The use of electronic-only journals in scientific research. *Issues in science and technology librarianship*. http://www.istl.org/02-summer/refereed.html

Lorie, Raymond (2002). *The UVC, a method for preserving digital documents : proof of concept*. No. 4 in IBM/KB Long-Term Preservation Study Report Series, Amsterdam: IBM Netherlands.
http://www.kb.nl/kb/ict/dea/ltp/reports/4-uvc.pdf

Lorie, Raymond A. (2000). The long term preservation of digital information. http://www.si.umich.edu/CAMILEON/Emulation

Lu, Shiyong, Ming Dong, and Farshad Fotouhi (2002). The Semantic Web: opportunities and challenges for next-generation Web applications. *Information research* 7. http://informationr.net/ir/7-4/paper134.html

Lynch, Clifford A. (2003). Institutional repositories: essential infrastructure For scholarship In the digital age. *Portal: libraries and the academy* 3(2):327–336, also published in ARL Bimonthly Report 226, February 2003 (http://www.arl.org/newsltr/226/ir.html).
http://muse.jhu.edu/journals/portal_libraries_and_the_academy/v003/3.2lynch.html

Macdonell, Karen L. (1999). The changing nature of the scholarly journal.
http://www.slais.ubc.ca/courses/libr500/fall1999/www_presentations/K_macdonell/index.html

Mackenzie Owen, J.S. (1987). Innovatie binnen de informatieketen. *I&I, Informatie en informatiebeleid* 5(4):15–17.

Mackenzie Owen, J.S. (1989a). Kennis, betoog en conversatie: veranderingen in het gebruik van de informatieketen [Knowledge, argument and conversation: changes in the use of the information chain] (with a commentary by J.M. van Oorschot). In *Sociaal-wetenschappelijke informatie en kennisvorming in onderzoek, onderzoeksbeleid en beroep*, A.F. Marks, ed., pp. 39–54, Amsterdam: SWIDOC.

Mackenzie Owen, J.S. (1989b). Technology, users and the information chain: changes in the communication of science. Paper presented at the Vienna Centre's 5th General Conference of ECSSID 'The role of social science information in knowledge creation', Berlin, January 22-24. *International journal of information & library research* 1(2):106–115.

Mackenzie Owen, J.S. (1996). Preservation of digital materials for libraries. *LIBER quarterly* 6(4):435–451.

Mackenzie Owen, J.S. (1998). Organizing for digital archiving: new distribution models in the scientific information chain. In *The impact of electronic publishing on the academic community*, I. Butterworth, ed., pp. 128–133, London: Portland Press.
http://cf.hum.uva.nl/bai/home/jmackenzie/pubs/Stockholm2.htm

Mackenzie Owen, J.S. (2002). The new dissemination of knowledge: digital libraries and institutional roles in scholarly publishing. *Journal of economic methodology* 9(3):275–288.
http://cf.hum.uva.nl/bai/home/jmackenzie/pubs/WESP2-jmo-artikel.pdf

Mackenzie Owen, J.S. (2003). E-publishing voor wetenschappers: digitaal vastleggen, publiceren en distribueren van onderzoeksinformatie. *Handboek informatiewetenschap* I540:1–32.
http://cf.hum.uva.nl/bai/home/jmackenzie/pubs/E-publishing.pdf

Mackenzie Owen, J.S. and J. van Halm (1987). *Effecten van technologie op de informatieketen: een onderzoek naar ontwikkelingen in de overdracht van wetenschappelijke, technische en bedrijfsgerichte informatie*. Den Haag: Bibliotheekraad.

Mackenzie Owen, J.S. and J. van Halm (1989). *Innovation in the information chain: the effects of technological development on the provision of scientific and technical information*. London, New York: Routledge, iSBN 0-415-03871-5.

Mackenzie Owen, J.S. and J. van der Walle (1996). *Deposit collections of electronic publications*. Luxembourg: European Commission.
http://cf.hum.uva.nl/bai/home/jmackenzie/pubs/ELDEP.zip

Mahajan, Vijay and Robert A. Peterson (1985). *Models for innovation diffusion*. No. 48 in Quantitative applications in the social sciences, Sage Publications.

Marron, H. (1971). Clearinghouses. In *Encyclopedia of library and information science*, vol. 5, A. Kent and H. Lancour, eds., pp. 196–197, New York: Marcel Dekker.

Martin, Mairéad, Grace Agnew, Davis L. Kuhlman, John H. McNair, William A. Rhodes, and Ron Tipton (2002). Federated digital rights management:

a proposed DRM solution for research and education. *D-lib magazine* 8(7/8). http://dlib.org/dlib/july02/martin/07martin.html

Mattlage, A. (1999). Networked scholarly publication. *Journal of academic librarianship* 25(4):313–321.

McClelland, J.F. (1994). *Technology Foresight 4: an information technology view*. London: Office of Science and Technology.

McGrath, William E. et al., (2002). Current theory in library and information science. *Library trends* 50(3):309–574.

McKechnie, Lynne E.F. and Karen E. Pettigrew (2002). Surveying the use of theory in library and Information science research: a disciplinary perspective. *Library trends* 30(3):406–417.

McKemmish, Sue (1997). Yesterday, today and tomorrow: a continuum of responsibility. In *Preserving Yesterday, Managing Today and Challenging Tomorrow: Proceedings 14th National Convention RMAA, 1997*, pp. 19– , Perth, Records Management Association of Australia, also published in 'Naar een nieuw paradigma in de archivistiek' (red. P.J. Horsman, F.C.J. Ketelaar and T.H.P.M. Thomassen), Jaarboek 1999 (´s-Gravenhage: Stichting Archiefpublicaties, 1999.

McKemmish, Sue (2001). Placing records continuum theory and practice. *Archival science* 1(4):333–359.

McKiernan, Gerry (2002). E is for everything: the extra-ordinary, evolutionary E-journal. *Serials librarian* 41:293–321.

Meadows, A. Jack (1998). *Communicating research.* Academic Press.

Mellor, Phil, Paul Wheatley, and Derek Sergeant (2002). Migration on Request : A practical technique for digital preservation. CAMiLEON Project. http://www.si.umich.edu/CAMILEON/reports/migreq.pdf

Menou, M J (1971). Information revolution or revolution for information. In *International Conference on Training for Information Work, Rome, 15-19 November.*

Merton, R.K. (1973). *The sociology of science: theoretical and empirical investigations*. University of Chicago Press.

MLA Ad Hoc Committee on the Future of Scholarly Publishing (2003). The future of scholarly publishing. *Journal of Scholarly publishing* 34(2):65–82, first published by MLA in Profession, 2002.

Mobley, E.E. (1998). Ruminations on the Sci-Tech Serials Crisis. *Issues in science and technology librarianship.* http://www.library.ucsb.edu/istl/98-fall/article4.html

Moles, Abraham (1958). *Théorie de l'information et perception esthétique.* Paris: Flammarion, also published in a translation by Joel E. Cohen as 'Information theory and esthetic perception', University of Illinois Press, 1966.

Morgan, Betty Trebell (1928). *Histoire du Journal des Scavants depuis 1665 jusqu'en 1701.* Paris.

Morrissey, Frances (2002). Introduction to a semiotic of scientific meaning, and its implications for access to scientific works on the web. *Cataloging and classification quarterly* 33(3/4):67–97.

Moulthrop, Stuart (1991). You Say You Want a Revolution? Hypertext and the Laws of Media. *Postmodern Culture* 1(3). http://muse.jhu.edu/journals/postmodern_culture/v001/1.3moulthrop.html

Nelson, R.R. (1987). *Understanding technical change as an evolutionary process.* Elsevier Science.

Nentwich, Michael (2003). *Cyberscience: research in the age of the Internet.* Vienna: Austrian Academy of Sciences Press.

Newman, Eric J. (1990). The emergence of European publishing. *Library acquisitions: practice and theory* 14(3):233–235.

Nielsen, Jakob (1997a). How users read on the web. *Alertbox: current issues in web usability* (October 1). http://www.useit.com/alertbox/9710a.html

Nielsen, Jakob (1997b). PDF: unfit for human consumption. *Alertbox: current issues in web usability* (July 14). http://www.useit.com/alertbox/20030714.html

Nielsen, Jakob (2001). Avoid PDF for on-screen reading. *Alertbox: current issues in web usability* (June 1). http://www.useit.com/alertbox/20010610.html

Nwaobi, Godwin Chukwudum (2001). *The new globalization era and digitalization debate: an economists perspective.* Economics Working Paper Archive at WUSTL. http://econpapers.hhs.se/paper/wpawuwpit/0112001.htm

Odelberg, W. (1978). Berzelius, father of scientific abstract and reference literature in the natural sciences. In *Knowledge and development. Reshaping library and information services for the world of tomorrow. A festschrift for*

Björn Tell, S. Schwarz and U. Willers, eds., pp. 99.11–27, Stockholm: Royal Institute of Technology.

Odlyzko, Andrew (1999). Competition and cooperation: libraries and publishers in the transition to electronic scholarly records, revised version, April 27. http://www.dtc.umn.edu/~odlyzko/doc/competition.cooperation.pdf

O'Hara, Kenton and Abigail Sellen (1997). A comparison of reading paper and on-line documents. In *CHI '97: Proceedings of the SIGCHI conference on Human factors in computing systems*, pp. 335–342, ACM Press.

Orlikowski, Wanda and JoAnne Yates (1998). Genre systems: structuring interaction through communicative norms.
http://ccs.mit.edu/papers/ccswp205

Orlikowski, Wanda J. and Debra C. Gash (1994). Technological frames: making sense of information technology in organizations. *ACM Transactions on information systems* 12(2):174–207.

Orlikowski, Wanda J. and Debra Carol Gash (1992). Changing frames: understanding technological change in organizations. Massachusetts Institute of Technology (MIT), Sloan School of Management Working papers number 3368-92., also available http://ideas.repec.org/p/mit/sloanp/2382.html.
https://dspace.mit.edu/bitstream/1721.1/2382/1/
SWP-3368-25121345-CISR-236.pdf

Ornstein, Martha (1913). *The role of the scientific societies in the seventeenth century*. New York.

Otlet, Paul (1934). *Traité de documentation: le livre sur le livre: théorie et pratique*. Buxelles: Editones Mundaneum.

OUP (2004). *About OUP - history*. Oxford University Press.
http://www.oup.com/about/history/

Parliamentary Office of Science and Technology (2002). Peer review. *Postnote* (182):1–4.
http://www.parliament.uk/post/pn182.pdf

Peacey, Allan (1996). The development of the tobacco pipe kiln in the British Isles. *Internet archaeology* 1(1).
http://intarch.ac.uk/journal/issue1/peacey_toc.html

Peek, Robin and Jeffrey Pomerantz (1998). The traditional scholarly publishers legitimize the web. *Journal of the American Society for Information Science* 49(11):983–989.

Penrose, Ann and Steven B. Katz (2004). *Writing in the sciences:exploring conventions of scientific discourse*. Longman.

Pera, Marcello (1994). *The discourses of science*. University of Chicago Press.

Petitcolas, Fabien A. P. and H. J. Kim (2003). *Digital watermarking: proceedings of the first International Workshop on Digital Watermarking, Seoul, Korea, 21-22 November 200*. No. 2613 in Lecture Notes in Computer Science, Berlin: Springer Verlag.

Pettigrew, Karen E. and Lynne McKechnie (2001). The use of theory in information science research. *Journal of the American Society for Information Science and Technology* 52(1):62–73.

Phelps, Th. A. (1998). *Multivalent documents: anytime, anywhere, any type, every way user-improvable digital documents and systems*. Ph.D. thesis. http://www.cs.berkeley.edu/helps/papers/dissertation-abstract.html

Picot, Arnold, ed. (2003). *Digital rights management*. Berlin: Springer.

Pinch, Trevor J. and Wiebe E. Bijker (1987). The social construction of facts and artifacts: or how the sociology of science and the sociology of technology might benefit each other. In *The social construction of technological systems: new directions in the sociology and history of technology*, Wiebe E. Bijker, Thomas P. Hughes, and Trevor Pinch, eds., pp. 17–50, MIT Press.

Polanyi, Michael (1969). *Knowing and being*. University of Chicago Press.

Popper, Karl R. (1972). *Objective knowledge: an evolutionary approach*. Oxford: Clarendon Press.

Popper, Karl R. (1978). Three worlds. The Tanner lecture on human values, delivered at the University of Michigan, April 7, 1978. http://www.tannerlectures.utah.edu/lectures/popper80.pdf

Poster, Mark (2001). *What's the matter with the Internet?* University of Minnesota Press.

Prelli, Lawrence J. (1989). *A rhetoric of science: inventing scientific discourse*. University of South Carolina Press.

Public Library of Science (2001). Open letter. *Science* 23(March):2318.

Pyenson, Lewis and Susan Sheets-Pyenson (1999). *Servants of nature: the Fontana history of scientific institutions, enterprises and sensibilities*. London: Fontana Press.

Raan, Anthony F.J. van (2001). Bibliometrics and internet: Some observations and expectations. *Scientometrics* 50(1):59–63, tY - JOUR.

Raber, Douglas and John M. Budd (2003). Information as sign: semiotics and information science. *Journal of documentation* 59(5):507–522.

Radford, Gary P. (2003). Trapped in our own discursive formations: an archaeology of library and information science. *Library quarterly* 73(1):1–18.

Raman, T.V. (1994). AsTeR: Audio system for technical readings. http://www.rit.edu/~easi/itd/itdv01n4/article2.htm

Ravetz, Jerome R. (1971). *Scientific knowledge and its social problems*. Oxford: Clarendon Press.

Reddy, Michael J. (1993). The conduit metaphor: a case of frame conflict in our language about language. In *Metaphor and thought*, Andrew Ortony, ed., pp. 284–297, Cambridge: Cambridge University Press.

Robins, Kevin and Frank Webster (1999). *Times of the technoculture*. London: Routledge.

Roes, Hans (1994). Electronic journals: a survey of the literature and the net. *Journal of information networking* 2(3):169–186. http://drcwww.kub.nl/oes/articles/ej_join.htm

Rogers, E.M. (2003). *Diffusion of innovations*. New York: Free Press, 5th ed.

Roosendaal, Hans E., Peter A. Th. M. Geurts, and Paul van der Vet (2001). Developments in scientific communication: considerations on the value chain. *Information services and use* 2001(1):13–32.

Roosendaal, Hans E. and Peter A.Th.M. Geurts (1997). Forces and functions in scientific communication: an analysis of their interplay. In *Proceedings of the Conference on Co-operative Research in Information Systems in Physics, University of Oldenburg, Germany, September 1-3*, M. Karttunen, K. Holmlund, and E.R. Hilf, eds., also available from http://www.ub.utwente.nl/webdocs/dinkel/1/00000001.pdf. http://www.physik.uni-oldenburg.de/conferences/crisp97/roosendaal.html

Rothenberg, Jeff (1999). Avoiding technological quicksand: finding a viable technical foundation for digital preservation. Council on Library and Information Resources. http://www.clir.org/pubs/reports/rothenberg/pub77.pdf

Rothenberg, Jeff (2000). *An experiment in using emulation to preserve digital publications*. NEDLIB Report Series ; 2, The Hague: Koninklijke Bibliotheek, checked on 2002-09-04. http://www.si.umich.edu/CAMILEON/Emulation

Rowland, F. (1997). Print journals: fit for the future? *Ariadne*. http://www.ukoln.ac.uk/ariadne/issue7/fytton/

Rowland, Fytton (2002). The peer review process. *Learned publishing* 15:247–258.

Rusch-Feja, Diann and Uta Siebeky (1999). Evaluation of Usage and Acceptance of Electronic Journals. Results of an Electronic Survey of Max Planck Society Researchers including Usage Statistics from Elsevier, Springer and Academic Press (Full Report). *D-lib magazine* 5(10). http://www.dlib.org/dlib/october99/rusch-feja/10rusch-feja-full-report.html

Ryle, Gilbert (1963). *The concept of mind*. Penguin Books, first published by Hutchinson in 1949.

Sapp, Gregg and Ron Gilmour (2002). A brief history of the future of academic libraries: predictions and speculations from the literature of the profession, 1975 to 2000 – part one, 1975 to 1989. *Portal: Libraries and the Academy* 2(4):553–576. http://muse.jhu.edu/journals/portal_libraries_and_the_academy/v002/2.4sapp.html

Sapp, Gregg and Ron Gilmour (2003). A brief history of the future of academic libraries: predictions and speculations from the literature of the profession, 1975 to 2000 – part two, 1990 to 2000. *Portal: Libraries and the Academy* 3(1):13–34. http://muse.jhu.edu/journals/portal_libraries_and_the_academy/v003/3.1gilmour.html

Savenije, Bas (2003). The FIGARO project: a new approach towards academic publishing. *Learned publishing* 16(3):183–188.

Schaffner, A.C. (1994). The future of scientific journals: lessons from the past. *Information technology and libraries* 13:239–249. http://www.msri.org/activities/events/9495/fmc/Schaffner.html

Schneier, Bruce (2001). The futility of digital copy protection. *Crypto-gram newsletter* (May 15). http://www.schneier.com/crypto-gram-0105.htm

Schofield, Helen (1999). The evolution of the secondary literature in chemistry. In *History of information science: proceedings of the 1998 Conference on the History and Heritage of Science Information Systems*, Mary Ellen Bowden, Trudi Bellardo Hahn, and Robert V. Williams, eds., Medford, NJ: Information Today. http://www.chemheritage.org/explore/ASIS_documents/ASIS98_Schofield.pdf

Schulenburger, David E. (1998). Scholarly communication and the need for collective action: a statement by the Chief Academic Officers of the Big 12. http://www.big12plus.org/pressreleases/scholar.html

Schulenburger, David E. (1999). Moving with dispatch to resolve the scholarly communication crisis: From here to NEAR. In *Confronting the challenges of the digital era – Proceedings of the 133rd Annual Meeting of the Association of Research Libraries.* http://www.arl.org/arl/proceedings/133/shulenburger.html

Schön, Donald A. (1963). Champions for radical new inventions. *Harvard business review* 41:77–86.

Shannon, Claude E. and Warren Weaver (1949). *The mathematical theory of information.* University of Illinois Press.

Shapin, Steven (1996). *The Scientific Revolution.* Chicago: Chicago University Press.

Shaw, Dennis and Roger Elliott (1998). Executive Summary Report. In *Proceedings of ICSU Press Workshop on Economics, real costs and benefits of electronic publishing in science - a technical study. Keble College, University of Oxford UK, 31 March to 2 April 1998.* http://www.bodley.ox.ac.uk/icsu/execsumm.htm

Shearer, M.K. (2003). Institutional repositories: towards the identification of critical success factors. *Canadian journal of information and library science* 27:89.

Shum, Simon Buckingham and Tamara Sumner (2001). JIME: An interactive journal for interactive media. *First Monday* 6(2). http://www.firstmonday.org/issues/issue6_2/buckingham_shum/index.html

Siler, JM (2000). From Gutenberg to Gateway: Electronic publishing at University Presses. *Journal of scholarly publishing* 32:9.

Singh, Ripudaman, Rostislav Chudoba, K. Gopal, and Carsten Koenke (1998). IMMJ: interactive multi-media journals in science and technology - prospects and issues. *Ejournal* 8(2). http://www.ucalgary.ca/ejournal/archive/v8/immj3.htm

Slevin, James (2000). *The Internet and society.* Polity Press.

Smiraglia, Richard P. (2002). The progress of theory in knowledge building. *Library trends* 50(3):330–349.

Smith, John W.T. (1999a). The deconstructed journal: a new model for academic publishing. *Learned publishing* 12:79–91.

Smith, Richard (1999b). Opening up BMJ peer review. *British medical journal* (318):4–5.
http://bmj.bmjjournals.com/cgi/content/full/318/7175/4

de Solla Price, Derek (1975). *Science since Babylon*. Yale University Press, enlarged edition ed.

Solomon, David J. (2002). Talking past each other: making sense of the debate over electronic publication. *First monday* 7.
http://www.firstmonday.dk/issues/issue7_8/solomon/index.html

Van de Sompel, Herbert et al., (2004). Rethinking scholarly communication: building the system that scholars deserve. *D-Lib Magazine* 10(9).
http://www.dlib.org/dlib/september04/vandesompel/09vandesompel.html

Sonneland, Helge M. (2001). Electronic publications, copyright and access. *Nordinfo-NYTT-4/2001*. http://www.nordinfo.helsinki.fi/publications/nordnytt/nnytt4_01/sonneland.htm

Sosteric, Mike, Yuwei Shi, and Olivier Wenker (2001). The upcoming revolution in the scholarly communication system. *Journal of electronic publishing* 7. http://www.press.umich.edu/jep/07-02/sosteric.html

Star, Susan Leigh and Karen Ruhleder (1996). Steps toward an ecology of infrastructure: design and access for large information spaces. *Information systems research* 7(1):111–134.

Stephens, Mitchell (1989). *A history of news*. Penguin.

Stout, David (1999). ICTs and technology foresight. In *Society on the line: information politics in the digital ages*, William H. Dutton, ed., pp. 333–335, Oxford University Press.

Strickland, Lee S. (2003a). Copyrights's dilemma today: fair use or unfair constraints? Part 1: the battle over file sharing. *Bulletin of the American Society for Information Science and Technology* 30(1):7–11.

Strickland, Lee S. (2003b). Copyrights's dilemma today: fair use or unfair constraints? Part 2: the DMCA, the TEACH Act and other e-copying considerations. *Bulletin of the American Society for Information Science and Technology* 30(2):18–23.

Subramanyam, K. (1979). Scientific literature. In *Encyclopedia of library and information science*, vol. 26, Allen Kent, Harold Lancour, and Jay E. Daily, eds., pp. 376–548, New York: Marcel Dekker.

Sullivan, D.L. (2000). Keeping the rhetoric orthodox: forum control in science. *Technical communication quarterly* 9(2):125–146.

Sumner, Tamara and Simon Buckingham Shum (1996). Open peer review & argumentation: loosening the paper chains on journals. *Ariadne* 5. http://www.ariadne.ac.uk/issue5/jime/

Sumner, Tamara and Simon Buckingham Shum (1998). From documents to discourse: shifting conceptions of scholarly publishing. In *CHI'98 : Human factors in computing systems, Los Angeles, 18–23 April 1998*, pp. 95–102. http://kmi.open.ac.uk/publications/papers/kmi-tr-50.pdf

Søndergaard, Trine Fjordback, Jack Andersen, and Birger Hjørland (2003). Documents and the communication of scientific and scholarly information: Revising and updating the UNISIST model. *Journal of documentation* 59(3):278–320.

Tananbaum, Greg (2003). Of wolves and boys: the scholarly communication crisis. *Learned publishing* 16(4):285–289.

Tashakkori, Abbas and Charles Teddlie (1998). *Mixed methodology: combining quantitative and qualitative approaches*. Thousand Oaks: Sage.

Task Force on Archiving of Digital Information (1996). Preserving digital Information: report of the Task Force on Archiving of Digital Information. Commission on Preservation and Access Research Libraries Group, Inc. http://www.clir.org/pubs/abstract/pub63.html

Templeton, M. (1994). Letter to the Editor. *Journal of the American Society for Information Science* 45(8):567.

Tenopir, Carol (2003). *Use and users of electronic library resources: an overview and analysis of recent research studies*. Washington: CLIR. http://www.clir.org/pubs/reports/pub120/pub120.pdf

Tenopir, Carol and Donald W. King (2000). *Towards electronic journals: realities for scientists, librarians and publishers*. SLA.

Teufel, Simone and Marc Moens (1999). Discourse-level argumentation in scientific articles: human and automatic annotation. In *ACL Workshop Towards Standards and Tools for Discourse Tagging*.

Thiel, Sh. (1998). The online newspaper: a postmodern medium. *Journal of electronic publishing* 4. http://www.press.umich.edu/jep/04-01/thiel.html

Thompson, John B. (1990). *Ideology and modern culture*. Polity Press.

Travers, Michael David (1996). *Programming with agents: new metaphors for thinking about computation.* Ph.D. thesis, Massachusetts Institute of Technology. http://xenia.media.mit.edu/~mt/thesis/mt-thesis-2.1.html\#Heading25

Treloar, A. (1999). Rethinking the library's role in publishing. *Learned publishing* 12:25–31.

Trosow, Samule E. (2001). Standpoint epistemology as an alternative methodology for library and information science. *Library quarterly* 71(3):360–382.

Turing, Alan (1950). Computing machinery and intelligence. *Mind* 59(236):433–460.

Turk, Ziga (2004). SciX: final report, sciX Deliverable D20. http://www.scix.net/d/20/d20-101.pdf

UNESCO/ICSU (1971). *Study report on the feasibility of a world science information system.* Paris: UNESCO.

Upward, Frank (1996). Structuring the records continuum - part one: postcustodial principles and properties. *Archives and manucscripts* 24(2). http://www.sims.monash.edu.au/research/rcrg/publications/recordscontinuum/fupp1.html

Upward, Frank (1997). Structuring the records continuum, part two: structuration theory and recordkeeping. *Archives and manucscripts* 25(1). http://www.sims.monash.edu.au/research/rcrg/publications/recordscontinuum/fupp2.html

Upward, Frank (2001). Modelling the continuum as paradigm shift in recordkeeping and archiving processes, and beyond: a personal reflection. *Records Management Journal.* http://www.sims.monash.edu.au/research/rcrg/publications/FrankURMJ2001.pdf

Uren, Voctoria, Simon Buckingham Shum, Gangmin Li, John Domingue, and Enrico Motta (2003). Scholarly publishing and argument in hyperspace. In *Proceedings of the twelfth international conference on World Wide Web, 2003.*, pp. 244–250.

Urquhart, D.J. (1948). The organization of the distribution of scientific and technical information. In *Royal Society Scientific Information Conference, 21 June - 2 July 1948: report and papers submitted*, pp. 524–527, London: The Royal Society.

Vickery, Brian (1997). Metatheory and information science. *Journal of documentation* 53(5):457–476.

Vickery, Brian C. (2000). *Scientific communication in history*. Lanham and London: Scarecrow Press.

Voorbij, Hendrik Jan (2005). *Van gedrukte naar elektronische tijdschriften: gevolgen voor publicatiemodellen en evaluatiemethoden*. Ph.D. thesis, Universiteit van Amsterdam.

W3C (website). Semantic Web.
http://www.w3.org/2001/sw/

W3C-Consortium (2001). XML Linking Language (XLink) Version 1.0. W3C Recommendation 27 June 2001.
http://www.w3.org/TR/2001/REC-xlink-20010627/

Wallis, John (1700). The origin of the Royal Society. From Account of some passages of his life.
http://www.fordham.edu/halsall/mod/1662royalsociety.html

Warner, Julian (2001a). *Information, knowledge, text*. Lanham: Scarecrow Press.

Warner, Julian (2001b). W(h)ither information science? *Library quarterly* 71(2):243–255.

Weaver, Warren (1949). Recent contributions to the mathematical theory of communication. In *Shannon & Weaver, The mathematical theory of communication*, chap. 1.2, pp. 4–6, University of Illinois Press.

Weller, A.C. (2001). *Editorial peer review: its strengths and weaknesses*. ASIST.

Werry, Chris (2001). The work of education in the age of E-college. *First monday* 6(5). http://firstmonday.org/issues/issue6_5/werry/index.html

Westfall, Richard S. (1986). The scientific revolution. *History of Science Society newsletter* 15(3).
http://web.clas.ufl.edu/users/rhatch/pages/03-Sci-Rev/SCI-REV-Home/05-RSW-Sci-Rev.htm

Westfall, Richard S. (1995). Sallo, Denys [Denis] de. Galileo Project, Rice University. http://es.rice.edu/ES/humsoc/Galileo/Catalog/Files/sallo.html

Wheatley, Paul (2001). Migration - a CAMiLEON discussion paper.
https://worktools.si.umich.edu/workspaces/vbansal/001.nsf/Resources/5DC8588CB48E87F685256A200015B227/\$FILE/migration.htm

Whitley, R. (2000). *The intellectual and social organization of the sciences.* 2nd ed.

Wiegand, Wayne A. (1999). Tunnel vision and blind spots: what the past tells us about the present: reflections on the twentieth-century history of American librarianship. *Library quarterly* 69(1):1–32.

Wildemuth, Barbara M. (1993). Post-positivist research: two examples of methodological pluralism. *Library quarterly* 63(4):450–468.

Wilensky, R. (2000). Digital library resources as a basis for collaborative work. *Journal of the American Society for Information Science* 51:228–245.

Williams, R. (1997). The social shaping of information and communications technologies. In *The social shaping of information superhighways: European and American roads to the information society*, H. Kubicek, W.H. Dutton, and R. Williams, eds., pp. 200–338, New York: St. Martin´s Press.

Williams, Robin and David Edge (1996). The social shaping of technology. *Research policy* 25(6):856–899.

Wilson, Tom (2002). Information science and research methods. *Kniznicná a informacná veda [Library and Information Science]* 19:63–71. http://informationr.net/tdw/publ/papers/slovak02.html

Woolgar, Steve (1999). Analytic scepticism. In *Society on the line: information politics in the digital age*, William H. Dutton, ed., pp. 335–337, Oxford University Press.

Wouters, Paul and Repke de Vries (2004). Formally citing the web. *Journal of the American Society for Information Science* 14:1250–1260.

Yeung, Tim Au (2004). *Digital Preservation for Museums: Recommendations.* Minister of Public Works and Government Services, Canada, commissioned by the Canadian Heritage Information Network. http://www.chin.gc.ca/English/Pdf/Digital_Content/Preservation_Recommendations/preservation_recommendations.pdf

Zhang, Yin (2001). Scholarly use of internet-based electronic resources: a survey report. *Journal of the American Society for Information Science and Technology J* 52(8):628–654.

Ziman, John (2001). *Real science: what it is, and what it means.* Cambridge University Press.

Zucker, L. (1986). Production of trust. *Research in organizational behavior* 8:53–111.

Zuckerman, H. and R.K. Merton (1971). Patterns of evaluation in science: institutionalization, structure and functions of the referee system. *Minerva* 9(1):66–100.

Index

Information Science and Knowledge Management

1. C.W. Choo, B. Detlor and D. Turnbull: *Web Work*. Information Seeking and Knowledge Work on the World Wide Web. 2000 ISBN 0-7923-6460-0
2. C.A. Bean and R. Green (eds.): *Relationships in the Organization of Knowledge*. 2001 ISBN 0-7923-6813-4
3. R. Green, C.A. Bean and S.H. Myaeng (eds.): *The Semantics of Relationships*. An Interdisciplinary Perspective. 2002 ISBN 1-4020-0568-7
4. M. Huysman and D. de Wit: *Knowledge Sharing in Practice*. 2002 ISBN 1-4020-0584-9
5. B. Detlor: *Towards Knowledge Portals*. From Human Issues to Intelligent Agents. 2004 ISBN 1-4020-2053-8
6. A. Spink and B.J. Jansen: *Web Search: Public Searching of the Web*. 2004 ISBN 1-4020-2268-9
7. R. Szostak: *Classifying Science*. Phenomena, Data, Theories, Method, Practice. 2004 ISBN 1-4020-3094-0
8. A. Spink and C. Cole (eds.): *New Directions in Human Information Behaviour*. 2006 ISBN 1-4020-3667-1
9. H.F. Moed: *Citation Analysis in Research Evaluation*. Towards a Critical, Accurate and Policy Relevant Bibliometrics. 2005 ISBN 1-4020-3713-9
10. D. Blair: *Wittgenstein, Language and Information*. 'Back to the Rough Ground!'. 2006 ISBN 1-4020-4112-8
11. J. Mackenzie Owen: *The Scientific Article in the Age of Digitization*. 2007 ISBN 1-4020-5335-5

MIX
Papier aus verantwortungsvollen Quellen
Paper from responsible sources
FSC® C105338

If you have any concerns about our products,
you can contact us on
ProductSafety@springernature.com

In case Publisher is established outside the EU,
the EU authorized representative is:
Springer Nature Customer Service Center GmbH
Europaplatz 3, 69115 Heidelberg, Germany

Printed by Libri Plureos GmbH
in Hamburg, Germany